Privacy and the Past

Critical Issues in Health and Medicine

Edited by Rima D. Apple, University of Wisconsin–Madison, and Janet Golden, Rutgers University, Camden

Growing criticism of the U.S. health care system is coming from consumers, politicians, the media, activists, and healthcare professionals. Critical Issues in Health and Medicine is a collection of books that explores these contemporary dilemmas from a variety of perspectives, among them political, legal, historical, sociological, and comparative, and with attention to crucial dimensions such as race, gender, ethnicity, sexuality, and culture.

For a list of titles in the series, see the last page of the book.

Privacy and the Past

Research, Law, Archives, Ethics

Susan C. Lawrence

Rutgers University Press

New Brunswick, New Jersey, and London

Library of Congress Cataloging-in-Publication Data

Names: Lawrence, Susan C., author.

Title: Privacy and the past : research, law, archives, ethics / Susan C. Lawrence.

Description: New Brunswick, New Jersey : Rutgers University Press, 2016. |
Series: Critical issues in health and medicine | Includes bibliographical references and index.

Identifiers: LCCN 2015032497| ISBN 9780813574363 (hardback) | ISBN 9780813574370 (e-book (epub)) | ISBN 9780813574387 (e-book (web pdf))

Subjects: LCSH: Privacy, Right of—United States. | History—Research—Law and legislation—United States. | Historians—Legal status, laws, etc.—United States. | BISAC: MEDICAL / History. | LAW / Privacy. | SCIENCE / History. | MEDICAL / Ethics.

Classification: LCC KF1263.H57 L39 2016 | DDC 342.7308/58—dc23

LC record available at http://lccn.loc.gov/2015032497

A British Cataloging-in-Publication record for this book is available from the British Library.

Visit our website: http://rutgerspress.rutgers.edu

Manufactured in the United States of America

For David

Contents

Contents

Acknowledgments

I had no plans to write this book. I was, in fact, quite happily working on a completely different project when the incident described in chapter 1 occurred. I thus first owe a great debt to Marilyn Olson and Angela Keysor, who were then both graduate students at the University of Iowa. While meeting to talk about their projects, our conversations kept returning to what it means to be sensitive about privacy for the dead. They encouraged me to figure it all out. I have not been able to do that, but their confidence that I had something to say has helped to keep me going.

Many others over the years that this book took form deserve appreciation for having listened to me and shared their thoughts and experiences. These exchanges shaped the deep background to this book, as I absorbed ideas and impressions well before I started the formal process of applying for Institutional Review Board approval to interview archivists and historians. (I even got to do that twice, at the University of Nebraska–Lincoln and then all over again at The Ohio State University.) More friends and colleagues at the annual meetings of the American Association for the History of Medicine contributed to the development of my work than I can possibly remember for individual thanks, so a collective one will have to do: thank you all. Members of the Privacy and Confidentiality Roundtable of the Society of American Archivists were wonderfully welcoming to me when I began to attend SAA annual meetings in hopes of connecting with archivists to talk with informally and then, eventually, to interview. Phoebe Evans Letocha deserves a special thank you here. Archivists rock.

When I finally had my IRB-approved consent forms, people were very generous with their time for interviews. Not all interviews made it explicitly into this book, but every one of them helped me to understand more about the very diverse ways that archivists and historians understand privacy and the ways that it affects the history we can hope to write. My heartfelt thanks to (in alphabetical order): Emily Abel, Cynthia Connolly, Christopher Crenner, Elena Danielson, Amy Fitch, Janet Golden, Mark Greene, Sydney Halpern, Robert Lilly, Ellen More, Stephen Novak, Michael Pahn, Chris Paton, Constance Putnam, Jacki Rand, John Rees, Leslie Reagan, Susan Reverby, Benjamin Schneider, Eric

Schneider, Michael Simonson, Judith Wiener, Kelly Wooten, and two archivists who asked not to be named.

For reading all or parts of the manuscript and helping to make it better, I sincerely thank Rima Apple, John Burnham, Elena Danielson, Mark Greene, Angela Keysor, Michael Lawrence, Marilyn Olson, George Paulson, Scott Podolsky, Robyn Warhol, and John Harley Warner. All remaining errors, misunderstandings, and misinterpretations are entirely my responsibility.

Finally, I thank my husband, David Manderscheid, for sustaining me with laughter.

Privacy and the Past

Chapter 1

Introduction

The Historians, the County, and the Dead

This book began with one of those discoveries that delight a historian's heart: finding old documents in a storeroom. It has become an exploration of historians' ethical obligations to our research subjects: the living, the dead, and the living who are constantly passing away. It queries the extent to which the living do and should control access to information about people as historical actors and as unwitting participants in past events. It questions who gets to decide what is revealed and what is kept hidden. It takes on laws and court cases; it tackles archives and archivists. It looks at how demands to maintain individual privacy both protect and erase the identities of people whose stories make up the historical record. I did not fully understand the power that privacy laws, claims, and expectations have on how we do history until a graduate student and I encountered them during the ordinary course of her research work. In what follows, I have tried to understand that power and the role that historians have in negotiating the boundaries between individual privacy and historical accuracy.

Cease and Desist!

Susan had encouraged Marilyn to work on a local history topic for her master's degree in history at the University of Iowa, preferably a topic that required delving into unpublished archival sources. Given Susan's field—the history of medicine—and Marilyn's interests—how Americans have dealt with poverty—an examination of the ways that local governments in Iowa managed the responsibility of providing medical assistance to those on county relief in

the nineteenth century seemed like an ideal topic. There are very few good
secondary sources on the history of poverty in the rural Midwest. Most of
our knowledge of nineteenth-century poor relief depends upon studies done
of Eastern cities, counties, and states, or of special populations, such as Civil
War veterans. Sending advanced students into the ever-present gaps in the historical literature carries risks, of course, but Susan was confident that Marilyn
would find something upon which to build a master's essay, if only in the min-
ute books of county boards of supervisors, which are accessible under Iowa's
open records laws.

Marilyn set out to find records, canvassing the nearest counties for materials dealing with poor relief between settlement of the territory (1837) and
World War I. A great deal of local history still sits in county and city court
houses, from property deeds and plat books to probate records and committee ledgers. Marilyn found fairly rich records for the Cedar County Board of
Supervisors in Tipton, the county seat, along with a helpful tip from one of
the office staff whose days are punctuated by questions from those stepping
up to the counter that separates the People from the Government. The staff
person recollected that some of the county poor farm records just might still
exist at what had been a new (1916) main house for the county poor farm
but was now a county residential care facility. Marilyn found the building a
short distance outside of town, surrounded by acres of the farmland that had
once provided the poor farm's residents with food for their tables and labor
for their hands.

The facility's administrator guided Marilyn to a storage cupboard in the
finished third-floor attic and produced the Poor Farm Register, a volume kept
from 1871 to 1916. Understanding it to be a public record, she had opened
the volume over the years to those researching their family histories, and she
welcomed Marilyn to use it for her research. Marilyn certainly did so, snapping
digital photographs of its pages and returning occasionally to check details
as she transferred the entries from 1871 to 1893 (twenty-two years was quite
enough to cover) into a database and ran into questions about the handwriting.
She was thrilled to have found the register because, used with the board of
supervisors' meeting minutes, probate records, newspaper stories, and census
schedules, it could help her to piece together something about the lives of those
who had needed county aid in the decades after the Civil War.

All seemed business as usual for a history graduate student until one day
when she was working in the third floor storage area. The elevator opened and
several people exited. They were as surprised to see Marilyn as she was to see

them. One of the group, alarmed to see Marilyn working on the old records, told her to immediately "cease and desist!" Marilyn explained what she was working on and gave the horrified woman, who turned out to be a county social worker, her name and contact information. Shortly thereafter, Marilyn received an email that began:

> I am the HIPAA Compliance Officer for Cedar County. The Cedar County Board of Supervisors and Cedar County Attorney has [sic] requested that I get in touch with you regarding the research that you are doing. Would it be possible to meet with you and look at the research information you have documented thus far?[1]

Few would delight in getting a message from someone with lots of capitals in her job title and contents that include the word "attorney." Marilyn and Susan were unhappy, but not entirely baffled. Marilyn, a practicing hospital pharmacist, and Susan, a faculty person with an appointment in the Program for Biomedical Ethics and Medical Humanities in the Carver College of Medicine at the University of Iowa, knew about HIPAA. Under the Health Insurance Portability and Accountability Act, the federal Department of Health and Human Services had issued rules governing the use of individuals' health information, rules that put teeth into the long-standing ethical guidelines that kept medical information confidential. But what did that have to do with a nineteenth-century register from a long-defunct county poor farm?

Quite a bit, it turned out. Marilyn and Susan met with the HIPAA compliance officer, the social worker who had discovered Marilyn in the third-floor storage area, the county attorney, and the president of the county board of supervisors. Because the old poor farm building was being used as a residential care facility and medical information was kept and communicated in electronic form, all documents on the property were protected by the HIPAA Privacy Rule. And the Privacy Rule, as Susan and Marilyn soon learned, was both retroactive and, at the time, perpetual. It did not matter how old documents were, or that all of the people named in them were long dead. It did not matter that the documents were not medical documents, items created by doctors or nurses or some other health care practitioner. It did not even matter if the documents were those of a quite different institution. If the documents contained what looked like health information, then they were covered by HIPAA, and could have been withheld from public access forever. In this case, as the HIPAA compliance officer explained to Marilyn in her first email,

I have recently gone through those books page by page and discovered that all of them contain sensitive personal health information. I understand some of them are from the early 1800's [sic] and have historical information however those books contain information regarding residents that have family still alive and well. I'm sure you understand the magnitude and impact that information could have on a family.[2]

At the meeting, both the HIPAA compliance officer and the social worker elaborated on their view that the register contained "sensitive personal health information" and agreed that Marilyn should never have been allowed to see it in the first place. If she had asked officials at the court house for permission to use it in her research, they would have said no. The social worker seemed, in fact, even more distressed by the poor farm stewards' social judgments over the years about the reasons that some entered the poor farm (one man was "lazy," another a "deadbeat") than she was by the entry notes about broken limbs, old age, and smallpox.

Susan came to this meeting prepared to argue calmly that even if the register were covered by the HIPAA Privacy Rule, Marilyn should have access to it for her research. Susan first tried to explain that the register was not anything like a medical record, because the "sensitive personal health information" in some of the notes was in words written by the poor farm steward, a man hired to run the farm, not a doctor. His notes about why a poor person came to live on the farm were his lay evaluations, not expert ones, that having smallpox, or being "sick," or with "no home" were all valid reasons for needing county care. Sure, some of his evaluations seem insensitive to twenty-first-century eyes, such as "insane" or "lazy," but couldn't these county employees understand that these were nineteenth-century categories, not current ones? (No.) That the fact that a steward wrote these words down did not make them true in the context of current medical knowledge? (No.) Being told such things by a history professor did not sway the compliance officer or social worker because, as Susan later realized, such historical niceties did not matter to them.

Susan had discovered, before the meeting, that there are provisions in the Privacy Rule for researchers to use HIPAA protected records, since otherwise clinical and public health research would grind to a halt. Thus armed, Susan next argued that some accommodation could be reached between the county and Marilyn so that Marilyn could use information from the Poor Farm Register in her master's paper and in publications. She had more success with this tactic, but not because she was particularly persuasive. The county officers knew that Marilyn had already recorded quite a bit of information from the register, and had done so in good faith. If they could not persuade Marilyn that revealing any of that information would seriously harm Cedar County citizens, and appeal to her conscience not to use it in her work, what could they do? If someone complained to Washington about a HIPAA violation, the care facility—and its administrator—could have been penalized if a Department of Health and Human Services (DHHS) investigator agreed that a violation had in fact occurred, but the DHHS could not have prosecuted Marilyn under the HIPAA regulations. At this point, the possibility that Marilyn might agree to restrictions on what she revealed in exchange for formal permission to use the register in her research seemed like a way to minimize the damage that they imagined might ensue if Marilyn made names and "sensitive" information public. The initial meeting ended with the decision that Susan and Marilyn had to present the matter to the full county board of supervisors, and so they ended up on the agenda for the board's next meeting.

The meeting with the board of supervisors taught Susan and Marilyn a great deal about the political winds that were blowing around the county offices over Marilyn's access to the register. The supervisors were not particularly interested in Marilyn's research project, and did not seem to care one way or another about what the stewards had written in any of the entries. They cared about the fact that the register revealed that former Cedar County citizens had had to go to the poor farm in the nineteenth century, that some of them had received insulting labels from the stewards, that a few were revealed as "insane," sick or disabled, and—most important of all—that some may have descendants who were voters in the county. The possibility that even one of those living citizens might blame the supervisors for letting a researcher from the university learn about an ancestor's misfortunes, and then write about them in a paper accessible to the public, was the overriding concern. Thus, when Marilyn and Susan told the supervisors that the names of poor farm residents were explicitly listed in the census schedules of 1880 (some identified as "insane" or otherwise "defective"), and that these were already public documents, one of the board said that no one could blame them for that; anyone upset about those indiscretions could complain to the federal government.

In the end, the board told the county attorney to prepare an agreement that Marilyn would sign saying that she would not reveal any of the real names of individuals who were listed in the Cedar County Poor Farm Register in her master's paper and any resulting publications. Marilyn and Susan agreed to this plan, in principle, so that Marilyn could carry on with her project with the reluctant blessing of county authorities. When the county attorney produced a document that included, among other provisions, a provision that he would have prior approval of anything that Marilyn wrote using information from the Poor Farm Register, Susan took the document to one of the attorneys in the University of Iowa's counsel office. The attorney rewrote it without the prior-approval clause and with a few other modifications. The Cedar County attorney accepted these changes and, after several weeks of confusion and concern, the matter was settled.

Or, so it seemed. Certainly Marilyn continued with her research, completed her master's paper, and received her degree. She used her research for several presentations, and completed an academic article.[3] In the agreement she signed, she promised to destroy any research materials containing the real names of the poor farm residents. To date, she has not done so. Not everything went into the article she published, so she just might need that raw material again. Susan, sidetracked by HIPAA and its implications for research in the history of medicine, spent months reading the Federal Register, Health and Human Services websites, law review articles, court cases, and various erudite discussions of research ethics, and wrote an article about all of that for her peers.[4] The episode over the Cedar County Poor Farm Register still nagged, however.

Privacy and the Past

This story of a relatively minor culture clash between professional historians and county officials raises serious issues about the meaning of "privacy" for those who are dead. To put it bluntly: to what extent do historians have an obligation to consider the feelings of possible—or even known—living relatives when they use the real names of dead individuals found in the unpublished records they use in their research? Irrespective of what laws do and do not allow, do historians have ethical responsibilities to the living that should trump their professional responsibility to provide accurate, documented information in their work? If they do—why? If they do—for what kinds of information? If they do—who decides?

Consider, again, Marilyn's use of information from the Cedar County Poor Farm Register in her master's essay and now published article. Marilyn gave every person whose name appeared in the register a pseudonym when she used their details in her work. So, because she had traced these individuals in the census schedules, when she cited her sources she provided references to the schedules on which the individuals' real names appeared. An enterprising researcher following in her footsteps could go to the census schedule and see pages of names, some of which are the names of people who had been residents of the poor farm. But there the trail goes cold. The link between the real enumerated residents of the county and someone who needed county help is broken.

To establish the validity of Marilyn's claims, the enterprising researcher would then need to go to the county administrative offices and ask to consult the register. Given our experience, the answer will be "no" in the foreseeable future, even though modifications to the HIPAA Privacy Rule in 2013 mean that "protected health information" is, as of this writing, now covered by HIPAA only for fifty years after a person's death. The fifty-year period is only a guideline, since "covered entities may continue to provide privacy protections to decedent information beyond the fifty-year period." Thus, permission to consult the Poor Farm Register depends entirely on the authority of county government officials. Even though Marilyn was able to use the data she had gathered under a data use agreement, the officials have no obligation to extend that option to other researchers. One way to ensure access for everyone would be for someone to take the matter before a judge with the claim that the Poor Farm Register is a public document subject to Iowa's open records law, and to have that judge so rule. Another way would be for the county privacy officer and other officials to decide that, in fact, the register does not contain protected health information. They then could release it to the county historical society, to the local history collection of the public library, or deposit it in the state archives.[5] Perhaps, when political winds have shifted, this will happen. Perhaps not.

The broken chain between the census and the Poor Farm Register that Marilyn had to leave in her reference notes violates my sense of duty to the historical record, to the commitment that any person should be able to trace a historian's claims back to the original sources and to verify the accuracy of the historian's use of information. For historians, this process is the repeatable experiment of the bench scientist. If the trail cannot be verified, the reliability of the information is weakened, if not made seriously suspect. A major ethical breach in historical research, as in other fields, is faking data, and historians get caught only when others follow their footnotes and discover that documents do not exist or that what the records say and what the historian said they said do not correspond.[6] In other words, unless other researchers can access the Cedar County Poor Farm Register, for all anyone might know, Marilyn could simply have made up the information that she says it contains.

There is much more to my disquiet, however, than the highly unlikely possibility that another researcher might question Marilyn's work and need to check that she used the Poor Farm Register responsibly. In our meetings, the Cedar County social worker and HIPAA compliance officer asserted that harm would come to living individuals if the names and stewards' comments in the register were available to the public. Why? What lies at the core of these claims about harms to the living from information about those dead for decades? We did not interrogate the county officers for their knowledge of specific harms to identifiable living people, so what follows are the possibilities that Marilyn and I came up with in our many discussions about this over the years. Having spent time with academics actively teaching ethical and legal perspectives on human subjects research, and having served on a Social and Behavioral Sciences Institutional Review Board (IRB), I first applied the criteria used when considering social science research projects (which often use interviews and questionnaires) involving living people: can disclosures from the research cause psychological or social harms beyond those experienced in everyday life? The presumption is that questions about sensitive topics (such as childhood sexual abuse, torture, battlefield experiences) outside of a therapeutic relationship could cause serious mental harm. Similarly, the release of certain kinds of information (such as sexual abuse, domestic violence, drug use, adultery, crimes, cheating, and alcoholic blackouts) about identifiable individuals could affect their jobs, relationships, and risks for legal action. All of these are harms that the researcher must minimize and have protections for before starting her research. How does the release of information from the Poor Farm Register fare from this perspective?

First, there is the stigma of poverty. To enter the poor farm meant that a person was indigent. The person had no income, no savings, no unencumbered land to sell. The person also had no family residing in the county whose duty it was to provide a home and sustenance. The law expected that children would take care of aging parents, just as parents were expected to care for adult children too disabled to be self-supporting. But the extent to which the county actually forced relatives to care for other family members is unclear, although the board of supervisors certainly tried in a few cases. Perhaps the idea that one's grandparents or great-grandparents abandoned their own parent to the poor farm 130 years ago could be distressing, or socially embarrassing if the subject of new gossip in the twenty-first century. There is even the remote chance that the fact that an ancestor had sunk so low as to "go on the county" could cause psychological distress and tarnish a person's current reputation in a small community.

Second, there is the specific information recorded with each person that the stewards gave in the column headed "condition when admitted" in the register. The stewards never wrote more than a few words, and not infrequently left the column blank. It was this column that contained the "sensitive personal health information" that disquieted the HIPAA compliance officer and the pejorative labels ("lazy," "deadbeat") that offended the social worker. Here were the nineteenth-century terms that grate on twenty-first-century sensibilities: "insane," "idiocy," "deaf, dumb & blind," "cripple," and "colored." Here was the "health information" that needed protection: "liver complaint," "frozen feet," "rheumatism," "debilitated," "chills & fever," "sick," "pregnant," "broken arm," "hurt by fall," "smallpox," "quinsy," "mumps," "palsy," "ague," "dropsy," "has fits," "consumption," "invalid," "lame knee," "encumbered with flesh," and "old age." Quite a few were simply "in need" or had "no home." The assumption, again, that psychological or social harm could come to descendants of such individuals invests the stewards' opinions and observations with a great deal of power. But power to do what, exactly?

In general discussions when the 2002 HIPAA Privacy Rule was drafted, discussed, revised, and then finalized, two main themes came up when strict privacy advocates called for protections on the medical records of the dead. The first centers on the common belief that medical information exchanged between a physician and her or his patient is to be kept confidential, in an ethical bond that extends back to antiquity, the Greek physician Hippocrates and the famous oath that bears his name. This is, quite simply, a sacred cow that is frequently paraded, but rarely scrutinized.[7] The primary argument for doctor–patient confidentiality is the essential need for the patient to be willing to tell his physician the absolute truth about symptoms and behaviors so that the practitioner can diagnose and treat illnesses correctly. If doctors reveal patients' secrets, then not only will those patients lose their trust in their doctors, but other people may also refuse to trust their physicians and therefore harm their health and, in the case of infectious diseases, the health of others. This ethical relationship obviously does not apply to the Cedar County Poor Farm Register, because this was in no way a record of the intimacies of doctor–patient confidences. Most of the observations the stewards made, in fact, were conditions that would have been visible to anyone.

The second theme is more apt. Medical information about the dead can directly affect the medical privacy of the living if—and only if—the deceased had a condition that could have been inherited by his offspring.[8] Those who argued for protecting decedents' records regularly voiced the fear that a living person might have his health insurance coverage cancelled or modified upon discovering that a parent, grandparent, uncle, aunt, cousin, or child had a disease known to "run in families." Such information might also lead to refusal of employment, or loss of an existing job. Certainly, as genetic testing expands and more diseases are linked to specific genetic markers, the possibility of genetic discrimination becomes more real, even with federal legislation that prohibits it.[9] Any disclosure that threatens health insurance or employment could be a major harm, and must be prevented by protecting health information. That some diseases, such as alcoholism or schizophrenia, might carry a heritable social stigma, even if they lack unmistakable genetic signatures, further strengthened the appeal to protect the medical records of the dead. Extending this protection to "forever" for all past records covered by HIPAA, as the federal government did in 2002, however, seriously overestimates the diagnostic capabilities and disease terminology of medicine before World War II.

Could the medical information revealed by the poor farm stewards lead to genetic discrimination of living descendants? No. To say that the nineteenth-century stewards' diagnoses lack twenty-first-century rigor is a massive understatement. To worry that a steward's note about "palsy," for instance, might refer to Huntington's disease, one of the few conditions with a well-defined genetic marker that, if present, always leads to disease in an adult, would be absurd. "Palsy" was a generic term for localized muscle weakness or paralysis, with or without tremors, and can be a symptom of a wide range of conditions. Most historians of medicine argue strongly against attempts at retrospective diagnosis, the effort to make a twenty-first-century medical diagnosis based on practitioners' case records prior to the twentieth century, largely because such notes focus on patients' verbal histories, perhaps with some physical examination, and do not contain the radiographic scans, blood chemistry reports, tissue biopsy results and all of the other data that physicians now use to identify a condition. The stewards' notes in the Poor Farm Register, in short, lack the power to harm a living person's access to health care or employment opportunities.

I am back, then, to the possible psychological and social harm to living people caused by disclosure of the poor farm stewards' notes about particular incoming residents, quite apart from any distress caused by the revelation of their poverty. I must hazard a guess here, and argue that the chance of psychological and/or social harm resulting from disclosure that an ancestor had a broken arm, frozen feet, mumps, smallpox, or any of the other illnesses or injuries listed in the register, is exceedingly close to zero. None of these carry any particular stigma, nor would they surprise anyone even vaguely familiar with life in nineteenth-century America. The powerful words, instead, are those I noted as pejorative ("lazy," "deadbeat") or that identify disabilities and mental illness. These might affect the living in one of two ways. First, sometimes what "runs in families" are beliefs about the heritability of "insanity" or "idiocy" or deadbeat-ness that somehow taint the "blood" of a family tree. Such beliefs were significantly heightened during the hey-day of eugenics movements in the United States from the 1890s through the 1940s, and they still circulate in popular culture.[10] Such beliefs, nevertheless, are not supported by twenty-first-century genetics and developmental biology. Even if there are genetic components to the risk of being born with an intellectual disability, for developing a mental illness, or being lazier than others, there are also pre-natal, environmental, psychological, and social components that are not biologically inherited.[11] The stewards' powerful words thus play on false beliefs about what it means to have "insanity," intellectual disabilities, or character traits "run" in a family, especially as the stewards' terms encompass such broad categories. Any distress a person feels upon learning, or having it become known, that a great-grandparent entered the poor farm because he or she was considered "insane" and therefore feeling more at risk for "insanity" or for passing on "insanity," for instance, thus comes not from the stewards' words, but from cultural myths that vastly oversimplify the complexities of human genetics and development.

The second way that such terms might affect the living returns us to the possible harm that having a "deadbeat" or "insane" ancestor causes for a person's psychological well-being or social reputation, quite apart from any suggestion that those conditions are heritable. I am neither a psychologist nor a sociologist, but I have searched the professional literature in both areas seeking information about psychological or social harms resulting from the revelation of ancestral misdeeds, illnesses, disabilities, sexual orientations, poverty, or other possibly stigmatized behavior or character traits. What appears repeatedly in the literature are the pain, guilt, and shame associated with disclosures of sexual and reproductive secrets (homosexuality, illegitimacy, mixed race, abortion, adoption, adultery, use of donor eggs or sperm), and sexual and/or physical abuse, among living parents, children, and siblings. Such disclosures can involve revelations about grandparents, cousins, aunts, uncles, and more distant kin, whether living or dead, when patterns of abuse or reproductive secrets repeat across generations or through extended families.[12] I have found nothing about any harms arising from disclosures of other issues—medical, financial, cultural, or social—involving deceased relatives when these are not directly linked to socially stigmatized sexual behavior. If learning about family secrets (apart from abuse or reproductive irregularities) is psychologically harmful, surely there would be discussions of it in the mainstream academic counseling literature. At the very least, if historians must restrict or disguise information about the dead because it can cause psychological harm to the living, those who advocate this stance should be obliged to support this claim with empirical evidence.[13]

An absence of studies obviously does not mean that disclosures of dead relatives' problems (again, apart from abuse and reproductive surprises) never bother anyone, only that such distress has not risen to a level that warrants special counseling or special protections. Genealogists do warn those starting research on their own families' histories that they may not like what they find scattered in the family tree, and that other relatives might be unhappy or upset about revealing criminal records, lies about military service, discrepancies between marriage and childbirth dates, or other less than savory facts. Quarrels erupt, feelings are hurt. Of course, what family historians choose to reveal to their family members can be kept relatively private, unless disclosed in publications or, as is increasingly the case, on websites.[14] Since genealogists have a stake in their families' past, however, they presumably make decisions about what to share in a public forum well aware that they then lose control over what other people make of it, including any negative judgments about their ancestors' occupations, ethnicities, religious affiliations, and a host of other social facts that could affect living kin.

Studies of reputation are similarly silent on any impact that the public revelation of new information about long-dead relatives has on the social standing of living people.[15] To affect the living, people in their communities would have to know that the names historians disclose were indeed those of relatives of the living individuals with whom they are actually acquainted. To cause harm to the living, moreover, the information would have to distress community feeling so much that what the dead had done, or who the dead were, worsened people's evaluation of living people's characters, social standing, financial probity, and other reputable attributes, to the extent that they would treat the exposed individuals differently. Does learning that a store owner, accountant, school teacher, or any other citizen had an "insane" great-grandparent at the poor farm change how that person is treated? Would people stop doing business with her? Disqualify him from teaching their children? If such disclosures mattered so much, surely examples would have crept into the sociological literature on reputation. Once again, an absence of studies does not mean that new disclosures about dead relatives never affect the reputations of living people, just that such events remain far under the radar of those whose business it is to research American social problems. And, to repeat again, if historians must restrict or disguise information about the dead because it can cause social harm to the living, those who advocate this stance should be obliged to support this claim with empirical evidence.

At this point, Marilyn usually reminds me that, in my lofty academic musings, I do not understand the culture of small towns and rural communities. She grew up in rural North Dakota. People know each other. They gossip. Busybodies thrive on talking about new tidbits about their neighbors, and their neighbors' relatives, that come to their attention. For people whose families have lived in the same area for generations, the passage of time does not obscure identities as it does for those in mobile metropolitan areas, or for people like me, who moved around a great deal as a child and whose relatives are scattered across the United States. When I have pressed her, she has patiently explained that the issue in tight-knit communities is not psychological or social harms that might come from revelations about ancestors; this isn't about the trauma of revelations about century-old poverty, or being avoided because people learn that great-grandpa was nuts. It is much simpler. It is about privacy. Maybe historians should be able to rummage as much as they want in old documents, but publishing the names of possible ancestors who went to the poor farm as deadbeats, insane, sick, or just broke means that community members might find out about them. And talk. And such unpleasant family details are just none of their business, no matter how long ago the events took place. It may well be that the Cedar County officers used the HIPAA provisions to prevent the names of Cedar County poor farm residents from becoming known in the community because, at a gut level, revealing them seemed to violate the privacy of living relatives, quite apart from any psychological or social harms that the officials imagined might come to them.

Invoking family privacy as a reason not to use the real names of long-dead people in historical work leads us into quite a different set of concerns. An analysis of our Cedar County adventure from the perspective of research ethics directed me towards what experts have to say about the medical, psychological, and social harms that disclosures can afflict on the living. An analysis of the same adventure from the perspective of the privacy rights of individuals enmeshed me in what experts have to say about American statute and common law concerning the privacy of the dead and their surviving relatives. In a nutshell, the dead themselves have no right to privacy. The dead have no rights at all, in fact, because they have ceased being persons.[16] For a privacy claim to work, as I explain in much more detail in chapter 3, someone has to have grounds to sue the person who violated his or her privacy. Since the dead cannot bring lawsuits, family members have to make a case that it is their privacy, not the privacy of the dead person, that has been compromised. In recent years, courts have ruled, in a few extraordinary cases, that surviving relatives have legitimately claimed that their privacy rights were violated when the media published photographs of the recent death scenes of their loved ones, especially photographs showing the dead body or body parts.[17] But the courts have certainly not established a right to family privacy that could possibly encompass long-dead or distantly connected relatives.

Appealing to the non-existent legal status of the privacy of the dead, of course, obviously may not assuage the descendants of Cedar County poor farm residents who object to gossip about a great-grandparent's poverty. Those who destroy the letters, diaries, and other manuscripts of their deceased relatives, friends, lovers, and partners vividly demonstrate their determination to ensure the deceased person's privacy and, presumably, their own, without regard for any later historical value such documents might have.[18] Historians can share this sensitivity to disclosing facts about the dead based on their own lived experiences in communities where ancestral deeds and attributes are still avidly discussed and judged. Applying this sensitivity in historical research and writing has consequences, however, consequences that all historians of the recent past must consider when deciding what to include or omit from their lectures, talks, papers, articles, and books.

For the Cedar County officials, disclosing the real names of individuals who had entered the poor farm in the nineteenth century was distasteful. Revealing their poverty, their illnesses and injuries, and the stewards' pejorative labels for some of them might expose living citizens to embarrassment or even shame. But should the historian buy into that embarrassment and shame? Do we really protect the living by agreeing to hide the identity of those who faced poverty, mental illness, or name-calling in a very different time, cloaking their names with false ones? If so, then the historian becomes complicit in perpetuating stigmas that she may actually want to try to reduce by bringing understanding and compassion to the complex lives of historical actors. Indeed, deciding to use pseudonyms makes the stigmas worse, because the historian has to behave as if it were true that living people could be harmed by the shame of learning, or having others learn, that an ancestor was once in so much need that she had to spend some time on the poor farm, or was labeled as insane or a deadbeat. So much has been done over the past sixty years to reduce, if not yet eliminate, the social censure surrounding intellectual and physical disabilities, poverty, and mental illness that to reinforce such prejudices by requiring the afflicted dead to be disguised surely makes these conditions worse for the living, not better. One of the central ways that historians contribute to the public good is to research and write about what they discover about the past as objectively as they can in order to dispel the myths that have supported prejudices, inequalities, discrimination, and distorted world views. To use false names for nineteenth-century poor house residents does not contribute to that goal when doing so appears to validate fears that using real names could embarrass or shame their descendants.

But isn't it better to sacrifice a few real names to protect the privacy of the living, even when what is being revealed are facts about long-dead relatives rather than about still-breathing people? Even when what is being protected are not recognized privacy rights, but are feelings that certain events, conditions, or attributes of long-dead family members should not be made available for public consumption? The problem with this view is that the real individuals who experienced poverty, insanity, disabilities, illness, and insults are then essentially erased from the historical record that Marilyn's hard work brought to light. Fake names make them anonymous, faceless data points, rather than acknowledged as specific individuals who were part of a human community. The real people behind the fake names stay hidden, moreover, both for those who might want them hidden and for those who might not. Some of the poor farm residents left traces in the public record because they had owned land, held public office, were mentioned in local newspapers, or appeared in probate proceedings, although no one will be allowed to learn of their connections to the poor farm through Marilyn's work. Other poor farm residents, however, did not. For them, the poor farm entry might be the only tangible evidence that they ever lived. In this case, the poorest members of society once again lose out, this time for good.

The idea that we should obscure the identities of people who died decades ago to protect possible descendants from the release of information that is "none of our business" implies that some living people should control access to knowledge about the past at the expense of the rest of us.[19] This approach belies the belief that the past belongs to all of us, warts and all. Who entered the Cedar County poor farm in the nineteenth century as a wanderer just passing through, or due to childless old age, or because mentally challenged and alone after his mother passed away is part of our collective past, whether or not we ever come close to Cedar County, Iowa. The past is fragile. In an age awash in information, it is easy to forget that so much of the past is forever inaccessible because documents, images, and things were lost, destroyed, forgotten, or just left to decay away. How many of the ninety or so poor farms in nineteenth-century Iowa even kept registers? Of those that did, how many registers have survived? Of those that have survived, how many are in any condition to be readable? Of those that are readable, how many have anyone to care about the fragments of lives they contain? Of those that have anyone to care, how many have had a professional-to-be historian study them and then publish an article that puts those lives in historical context? The answer is: one. That makes what Marilyn discovered everyone's business. Her work informs us about how nineteenth-century Iowans, who were nineteenth-century Americans, who were nineteenth-century human beings, took care of those in need under a particular system of laws, taxes, and community values. It is part of that rich network of everyday lives and ordinary decisions that, in constant flux, forms the present from the past, and thus touches everyone, if ever so lightly.

Arguing that some facts and identities of that past are "none of our business" thus edges close to advocating for a kind of censorship, whether imposed by county officials, archivists, and family members, or self-inflicted by the historian as she writes. There may be good reasons for destroying, suppressing, disguising, or sanitizing some of the past. Certainly plenty of people have done so, in the name of privacy, honor, hero-worship, patriotism, and national security, not to mention others' attempts to cover up individual and collective stupidities, sins, crimes, and unethical acts. Yet, if there are good reasons, we need to be very clear about what they might be and why agreeing to them would be more important than scrupulous honesty and openness about the past.

Guide to this Book

From talking with historians, archivists, graduate students, and others in the years since Marilyn and I had our adventure in Cedar County, it became very clear to me that conditional access to archival materials, confusion about privacy claims for the dead, and the creeping influence of the language of research ethics are all affecting those who work on the history of the last 200 years.[20] Anyone whose research area takes him to documents and images that contain information about the personal lives of individuals runs the chance of discovering details that some living person might find offensive, scandalous, private, or simply none of the historian's business. Indeed, it is all too easy to imagine a hypothetical critic challenging disclosures with a sense of high moral purpose and a vision of respect for the dead. Since historians, almost by definition, thrive on reading other people's letters, diaries, and account books—the more personal, the better—as a group we tend to resent the implication that we are simply nosy. We seek the details in personal papers in order to illuminate larger themes, from gender relations to the emotional context of political decision-making. (At this point, I am distinguishing historians from biographers, although their interests and methods overlap somewhat.)

Non-historians, especially those who think that most history is boring and a waste of time, understandably have little patience with our desire to delve into past lives. For them, supporting restrictions on how historians can access, read, and publish information about the dead can seem eminently reasonable. When protecting the privacy of decedents is interwoven with the language of the ethical obligation to protect the privacy of living research subjects, more-over, supporting restrictions is not just reasonable, it appears to be the right thing to do. There are many others whose perspectives are much more com-plicated than a simplistic division of historians from non-historians. Journal-ists, attorneys, archivists, physician-historians, philosophers, sociologists, and genealogists are among those who spend time with old materials. Each of them comes to the study of the past shaped by the norms and methodologies of their professions and avocations, and hence by expectations for what it means to respect the privacy of those with whom they interact, whether in person or via texts, sounds, and images. Indeed, historians do not, and will not, agree with each other about revealing sensitive information about named individuals.

This book engages the thorny questions surrounding privacy and histori-cal research in the United States.[21] The blow-by-blow analysis of the encoun-ter with Cedar County officials over Marilyn's use of the Poor Farm Register offered in this chapter illustrates why historians—and others—need to under-stand how privacy protections for the living have implications for historical research on both the living and the dead. Privacy concerns affect historical materials and historians' use of them at multiple points. First, there are ques-tions of survival and access. People destroy their own or their family mem-ber's private papers (or, more recently, media), entirely removing them from potential use; government employees shred or erase pesky records that are claimed to be private items. Documents may be kept, but saved in private hands, put on shelves in office storage rooms, or stored in electronic limbo, where no historian can examine them because few people even know they exist. Private and government records may go to a repository, but be closed to use for varying periods of time to protect individuals' privacy. Research-ers may be able to get access to closed or restricted records, perhaps, if they go through a vetting process set up by an archive, a Privacy Board, or an IRB (more on this in later chapters). Such records may have names and other identifying information redacted before the researcher even sees them, to ensure that no individual's identity escapes.

Once a historian has materials in front of him, a different set of considerations begins. He may see all of a collection's contents, but only after signing an agreement not to use names, or to omit names and other individual details. The historian thus chooses to agree to terms that others set. Finally, he may see all of a collection's items in full, and then be faced with ethical decisions about what to do about identifying people. How he chooses depends not only on his prior ethical sensibilities, but also on what he may think he knows about the law (can he be sued?) and about expectations for ethical behavior in the academy (are these people living? dead? related to sensitive people?).

Because so many misunderstandings arise from the ways that expectations about the ethical treatment of living research subjects spill over into expectations about the ethical treatment of the dead and their living descendants, in chapter 2 I outline key points in research ethics that can affect historians and the range of federal laws that protect of various kinds of information (e.g. educational, financial). In chapter 3, I discuss how the courts have dealt with privacy in contexts relevant to historians, including the single case I have discovered in which a historian was actually sued for a privacy violation. In chapter 4, I turn to archivists, who control access to records, and examine the privacy issues that they consider when acquiring records and allowing researchers to use them. In chapter 5, I discuss decisions that historians have made to conceal identities that they believed needed to be protected (either by not mentioning them at all in their work, by giving no names, or by using pseudonyms), or to disclose private information about named individuals that some might find objectionable.

I conclude, in chapter 6, with a call to push back against any regulatory language that deliberately or inadvertently protects the privacy of the dead. I appeal to historians to make self-censorship decisions with great care, to weigh the benefits and harms to the living along with their duty to the historical record and to shared knowledge. There are no formulas for making these decisions, and there should not be, given the vast diversity of historical materials and problems we take on. We can, nevertheless, do our best to make informed choices and to be able to defend those choices in the face of criticism that we do not understand that some things are neither our business nor ours to reveal.

Chapter 2

Research, Privacy, and Federal Regulations

Historians seeking legal guidance on how to deal with privacy issues when they research and write about the past can find it in four overlapping areas of legislative, regulatory, and judicial action: federal regulations for human subjects research protection; the range of federal laws that constrain disclosures of certain kind of information (educational, financial, medical); the First Amendment and journalism; and the civil laws that provide redress from harms, specifically tort law covering invasions of privacy. What all of these have in common is the language of the law: regulatory law issued to turn statutory mandates into working policies and procedures (such as the HIPAA Privacy Rule), statutes passed by Congress (such as the Federal Educational Rights and Privacy Act [FERPA]), legal principles established in the U.S. Constitution (First Amendment), and the traditions of common law (interpretations determined by the courts through the precedents of case law, in this case the law of torts).

Except for the regulations about human subjects protections, research is rarely mentioned in any of these areas. Yet all of them involve access to information about individuals and what happens when that information goes public. From that perspective, they all offer some insights into what historians may do to get and to publish details about living and dead individuals. This chapter focuses on statutory and regulatory laws formed to make some kinds of information open and to keep other kinds of information closed. It also covers a handful of instances where the courts have dealt with challenges to these laws in ways important to historians. The next chapter takes up the ways that courts have dealt with the constitutional issues (First Amendment) and civil actions against those who, the plaintiffs argued, infringed on their privacy by publishing facts about them.

Nothing in this chapter offers legal advice for specific situations, of course; the daunting technicalities of the law require expert guidance on a case by case basis, especially when state laws come into play alongside federal ones. Academics working in universities, moreover, need to pay attention to how their particular institutions interpret the federal regulatory language covering research with human beings. This chapter aims, instead, to lay out the range of approaches to privacy embedded in American laws and freedoms so that historians can arm themselves with what "privacy" means in legal terms as they make choices about how to deal with information in their work.

Research and the Prevention of Harm

Federal regulations mandate how research using human subjects may be performed. Until the early 1960s, investigators in all disciplines made up their own minds about how to treat the people who volunteered for research studies, with due regard for their own field's explicit or implicit standards for appropriate conduct. In the 1960s, thoughtful investigators and attorneys at the Clinical Center of the National Institutes of Health decided that research there, especially biomedical research on healthy subjects, needed to be reviewed by a committee of peers in order to ensure that basic ethical standards, such as informed consent, had been adequately considered by investigators.[1] After the revelations in the late 1960s and early 1970s that the subjects in various biomedical and psychological research projects had suffered coercion, lack of informed consent, and physical and psychological harms, the demand for federal oversight of all biomedical and behavioral research swept through Congress. The National Research Act of 1974 enabled the government to create administrative law to regulate all human subjects research in institutions that accept federal research funding. Today, the individual investigator subject to the regulations no longer has the authority to conduct research without prior review of her protocol and consent documents. The Office for Human Research Protections (OHRP), which is part of the U.S. Department of Health and Human Services (DHHS), has the responsibility for writing and enforcing these regulations. Most universities have voluntarily signed on to an expansive agreement with OHRP to ensure that all human subjects research carried out by their employees also follow the federal regulations, whatever the funding sources for the work.[2] The primary local mechanism for overseeing human subjects research is the Institutional

Review Board (IRB), either managed through an institution's office or run as an independent, for-profit enterprise.[3]

There are basic core requirements for an IRB that are usually overseen by an institution's research office, although the DHHS is the ultimate supervising authority. An IRB must have at least five members, with at least one scientist and one non-scientist, who have backgrounds appropriate for the research being reviewed. "Appropriate" can be taken broadly. While there are no accessible statistics on the actual composition of IRBs, it is safe to say that no IRB has members that represent all of the specific disciplines in which research is taking place. Due to the different orientation of the biomedical sciences, particularly those involving clinical and therapeutic interventions, most universities have multiple IRBs, with at least one for biomedical research and one for social and behavioral investigations. One member of any IRB must be a member of the community, moreover, with no affiliation to the home institution or to a person who is connected with the home institution. IRBs are encouraged to have members who can speak for particular vulnerable populations, such as children, prisoners, or disabled persons, especially when research on such populations is common at the institution. Finally, there are strict guidelines about conflicts of interest, so that IRB members do not evaluate research projects that in any way concern their own, their friends', or their relatives' research agendas.[4]

In 1978, the National Commission for the Protection of Human Subjects of Biomedical and Behavioral Research produced the *Belmont Report: Ethical Principles and Guidelines for the Protection of Human Subjects of Research*.[5] This document enshrines the principles of beneficence (do good), justice (do not discriminate among subjects; equalize risks and benefits) and autonomy (respect people as decision makers) as the bedrocks of ethical research methods, and its tenets were incorporated into the official set of federal regulations in the 1980s, now generally known as the Common Rule (45 CFR 46).[6] Before getting to the principles, however, the authors realized that they had to define research. One of their primary concerns, given the context of biomedical and behavioral work, was to distinguish research from ordinary medical practice or clinical therapy. So, the authors stated:

> By contrast [to practice or therapy], the term "research" designates an activity designed to test an hypothesis, permit conclusions to be drawn, and thereby to develop or contribute to generalizable knowledge (expressed, for example, in theories, principles, and statements of relationships). Research is usually described in a formal protocol that sets

forth an objective and a set of procedures designed to reach that objective.[7]

Do historians do research? We certainly believe that our investigations constitute research in the way that the word is used in ordinary speech. Yet we rarely test hypotheses in any formal sense, nor do most of us have a set of procedures that consists of much more than "go to libraries and read as many relevant secondary and printed primary sources as possible; go to archives and seek out relevant unpublished sources. Then try to make sense of it all using insights from theories developed in a range of other fields, from economics to feminist studies." We seek explanations of the particular, not universal theories or principles. Our explanations may involve "statements of relationships," but these are rarely relationships in any statistical or generalizable sense.

The belief that history can be made into a science, with the induction of "laws of history" out of innumerable particular facts, vanished in the 1960s, with the publication of the last volumes of Arnold Toynbee's *A Study of History*. His was the final attempt to discern fundamental, overarching (and untestable) generalizations that explained the course of world history from prehistoric times through the first half of the twentieth century.[8] Certainly a few brave authors, notably Jared Diamond (*Guns, Germs and Steel: The Fates of Human Societies* [1997], *Collapse: How Societies Fail or Succeed* [2004]), still urge historians to test hypotheses about the effect of material conditions and social structures on historical change by rigorous comparative analysis of isolated cultures. Diamond, not surprisingly, wrote his sweeping texts using the work of historians, but is himself a geographer and physiologist, so perhaps it is only non-historians who expect that history can rise to the "generalizable" knowledge of the natural, biomedical, and behavioral sciences.

Unfortunately, the *Belmont Report*'s authors did not place judicious adjectives in front of their use of "research." Could they not have written "biomedical research"? Or "biomedical and behavioral research"?[9] Instead, they drew a perfectly good word entirely into the domain of the biomedical and behavioral sciences and so unwittingly drew the research that historians do under their umbrella. For most humanists, the *Belmont Report*'s definition of research seems irrelevant, just unfortunate rhetoric that we can laugh off. Indeed, for most historians, it has been irrelevant, because the primary use of the *Belmont Report*'s definition has been in the federal rules guiding human subjects research and—as has been made crystal clear in the *Report* and rules—a human subject is, by definition, a living human being "about whom an investigator conducting research obtains (1) data through intervention or interaction with the individual [i.e. talking], or (2) identifiable private information."[10]

For historians who wish to talk to living people for research purposes, however, this definition is far from a laughing matter. As Zachary Schrag has detailed in *Ethical Imperialism: Institutional Review Boards and the Social Sciences, 1965–2001*, during the last three decades OHRP has explicitly decided that doing oral history may constitute human subjects research, because it might be a method in a project aimed to create "generalizable knowledge."[11] But, if some oral history projects are designed to lead to "generalizable results" and others are not, who determines which ones count as research? The answer to this question is complicated by the language of the federal guidelines: some categories of research are "exempt" from review, including:

> Research involving the use of educational tests (cognitive, diagnostic, aptitude, achievement), survey procedures, interview procedures or observation of public behavior, unless:
>
> (i)   information obtained is recorded in such a manner that human subjects can be identified, directly or through identifiers linked to the subjects; *and*
>
> (ii) any disclosure of the human subjects' responses outside the research could reasonably place the subjects at risk of criminal or civil liability or be damaging to the subjects' financial standing, employability, or reputation.[12]

Oral history typically does, indeed, include "identifiers" along with the audio recording and/or transcript of the interview, because the value of the interview for historical research is enhanced by knowing exactly who was speaking. The more problematic clause (ii) concerns the estimate of harms that might befall an identified interviewee if, for instance, she confessed to crimes, admitted wrongdoing, or disclosed unsavory facts about herself, even if she were to do so freely. So, not only is the question at hand whether an oral history project is "research," but also if it is research during which an identified person might "reasonably" do harm to herself by talking to a historian without ensuring that her identity is kept confidential. Is a project not research, exempt research, or research that needs to be reviewed? Again, who determines this?

No single answer exists, because it is the responsibility of each local IRB to interpret the federal guidelines and policy statements, and to set up procedures for reviewing researchers' protocols to ensure that those guidelines and policies are followed. IRBs are free to develop their own policies and procedures under the umbrella of the federal regulations, so they are by no means uniform, cookie-cutter bodies.[13] Two examples are quite enough to illustrate how this local flexibility affects oral historians (see table 1). At UCLA, oral historians may not decide that their projects are neither research nor exempt research. Only the UCLA Office of the Human Research Protection Program may determine the project's categorization, and so the historian must submit details about his work in order to receive a "certification of exemption for UCLA IRB review of oral history," if appropriate.[14] As Linda Shopes, former President of the Oral History Association, put it, "we find ourselves in the position of submitting a description of our oral history research protocols to an IRB in order to apply for an exemption from applying to the IRB."[15] Even if a historian is not seeking "generalizable knowledge" from his project, but intends to deposit the tapes and/or transcripts of his interviews of identifiable people in an archive for others to use, UCLA's IRB requires review because the deposited data may be used to contribute to "generalizable knowledge" at some point in the future.[16]

The University of Nebraska–Lincoln, in contrast, lets the historian decide if her interviews are intended to add to "generalizable knowledge." If they are not, then the historian need not submit his forms for prior review.[17] The work isn't "research" and 45 CFR 46 does not apply at all. Since historians have to seek permission from the interviewee in order to deposit any audio recordings or transcripts in an archive, including restrictions an interviewee puts on the materials, such as not explicitly identifying the speaker or asking that the tapes and transcripts be closed until after the interviewee's death, the UNL IRB has determined that the interviewee has already agreed that the materials may enter a repository open to the public, and so whatever a researcher uses them for in the future does not require prior IRB approval.

If the oral historian does not seek IRB approval that her work is exempt from human subjects review, however, then woe to her if a master's or doctoral thesis committee, journal editor, or book publisher requires documentation that the research underwent IRB scrutiny. It is completely against the federal regulations to vet a research protocol and consent forms after the fact, so IRBs will flatly refuse to evaluate a project retroactively.[18] The research may not be published and, since graduate theses are technically published, degrees may be denied. If a complaint is made to an IRB office about a faculty member conducting sensitive oral history, his research might be stopped, pending an inquiry.[19] If found in violation of IRB rules, she may be required to destroy all data collected so far, lose her university or federal research funding (if she had any), and could even be blacklisted from ever receiving funding again. She might be raked over the coals by a department chair, college dean, or higher administrator. She might even be brought up on ethics charges before a faculty judicial board, depending on the university's ethics policies.[20]

Table 1 Oral History Projects, Interviews, and IRBs

Policy: oral history not generalizable knowledge. • Individuals have unique stories. • Not research	45 CRF 46 does not apply	The oral historian can ask what she pleases. The interviewee has to give consent for the interview to be deposited in an archive, so the interviewee knows perfectly well that what he or she says is being recorded and will be accessible to the public, either immediately or after his or her death.	The IRB office never sees the project.
Policy: oral history may be generalizable knowledge. • Individuals *may be* seen as representing types or classes or groups of people. • Gathering data for a research repository—might be used for generalizable knowledge.	45 CFR 46 does apply	The oral historian plans to identify his interviewees (with their permission) but *won't* ask any questions about sensitive topics that could cause the interviewees legal, financial, or reputational harm if disclosed.	IRB office determines that the research project is exempt from IRB review.
Policy: oral history may be generalizable knowledge. • Individuals *may be* seen as representing types or classes or groups of people. • Gathering data for a research repository—might be used for generalizable knowledge.	45 CFR 46 does apply	The oral historian plans to identify her interviewees (with their permission) and *will* ask questions about sensitive topics that could cause them legal, financial, or reputational harm if disclosed.	IRB office determines that the research project *is not exempt, but may be expedited* (IRB chair review only) if the risks of harm are minimal. If more than minimal risk, the project must go to a full board.

There is a great deal of accumulated wisdom by oral historians for those seeking practical recommendations on how to develop and implement an oral history project.[21] The emerging collective advice leans towards telling university-based students and faculty to run all oral history projects by an IRB, and so to get a formal certification of exemption unless the university has a clear policy excluding oral history from review.[22] Taking this path—which is not a choice at some universities—opens up challenges for oral historians when the staff person doing the IRB pre-review decides that an oral history project is not, in fact, exempt. If there is any question that identified interviewees might be asked to disclose information that could "reasonably place the subjects at risk of criminal or civil liability or be damaging to the subjects' financial standing, employability, or reputation" then the staff person has to send it to the IRB chair. The chair or the chair's designee, in turn, can approve the project (this is called an "expedited review") or send it to the full IRB to consider. The IRB chair may determine that, because the historian is asking questions about long-past events, the chances that the interviewees will disclose information that could damage them in the present are very small. The chair decides, then, that the protocol is "minimal risk" and so reviews it herself for proper consent documents. If the chair decides, however, that interviewees likely might tell the oral historian personal information that could open them to legal problems, or harm their standing in some way, or cause psychological distress during the recall of difficult memories, then the project may be of more than "minimal risk" and so require a full board to consider.

Here is a serious opportunity for a clash of cultures. If an IRB chair and IRB members are not familiar with oral history but come instead from sociology, anthropology, and psychology (assuming a social/behavioral science board, not a biomedical one), then the intent to identify interviewees flies in the face of commonly accepted practices in social science research, where identities are routinely suppressed or masked.[23] One of the *Belmont Report*'s primary ethical principles is "respect for persons." In the world of human subjects protections, "respect for persons" entails protecting subjects' privacy ("sensitive research information [is] not linked to individuals") and confidentiality ("sensitive, identifiable information is collected for research purposes but access to the information is limited to a well-defined group—like the research team and the IRB").[24] Because the social sciences seek "generalizable knowledge," just like the biomedical sciences do, the uniqueness of the individual is far less important and interesting than the role the person has as a member of a class, a type, or a group. The less the subject is identified as a unique human being, in fact, the more the interview data can be used to "generalize" from the individual to a group. When the interviewee becomes one of a sample of subjects, then there is no reason not to de-identify him; the researcher seeks the characteristics of the group, not the experiences of a unique individual. Here, I believe, the routine methodology of the social sciences has been so intertwined with the ethical expectation that protecting privacy and confidentiality mean "respect for persons" that revealing identities seems unethical, in and of itself.[25]

The culture clash becomes even more intense when an IRB chair or IRB members have to consider the possible psychological, economic, legal, social, and reputational harms that might come from participating in the research study.[26] An interview might create psychological distress simply as the interviewee remembers hard times and difficult events, distress the person would not be feeling at that moment if not for the research project. Asking very personal questions about matters considered private, such as sexual behavior or drinking habits, may feel like an invasion of privacy, even when the person being interviewed is free to refuse to answer the questions. The public revelation of past decisions and behaviors might, quite without either the interviewer or interviewee envisioning it, harm the interviewee's position in the community, even, perhaps, have a negative effect on the interviewee's job, insurance coverage, relationship with law enforcement, family dynamics, friendships, or respectability.

For an IRB, the duty of the researcher is to minimize risks, and the IRB must weigh those risks in light of the possible good that the research can bring to the research subject (in therapeutic studies) or, more often, to the good of all people, to society or to those in particular situations (such as prisoners).[27] A high probability of serious harm makes a procedure high risk, and so requires a high likelihood of benefit, as when a new cancer drug is offered to those who have not had results from standard treatments, or individuals divulge their participation in serious criminal acts in a study of the effectiveness of community intervention programs to reduce crime. Relatively low benefit studies, as when an exercise physiologist seeks to evaluate the effect that different leg lifts have on the strength of a thigh muscle in people considered overweight, must have equally small harms, such as muscle soreness and the possible embarrassment of being weighed. The probability of those effects may well be high—many people might experience the soreness and embarrassment—but the harm is so low that the overall risk is considered low.[28]

As much as historians might value the importance of oral histories as vital additions to our understanding of the past, it is hard to argue that intellectual benefit to social scientists who are more used to appreciating the significance that psychological and sociological studies have for understanding living people right now. Since oral history is not likely to directly save lives, revise public policies on how to deal with drug addiction, or lead to ways to end bullying (among many possible examples), unless the historian makes an eloquent case for the benefits of her study to society, any risks to human subjects should be very low if not, indeed, "minimal."

According to the Common Rule, "minimal risk means that the probability and magnitude of harm or discomfort anticipated in the research are not greater in and of themselves than those ordinarily encountered in daily life or during the performance of routine physical or psychological examinations or tests."[29] So, if a historian intends to ask about sensitive topics that might cause psychological distress or harm the interviewee's standing in the community more than the interviewee could be expected to experience in daily life, an IRB could well ask the historian for serious changes to the study. For psychological distress, the historian might need to provide her interviewee with information about local resources for professional counseling. For potentially harmful disclosures, the historian simply might be told that she may not reveal her interviewees' identities. In that case, she must have security measures in place to protect those identities during her research, must completely mask identities if interview transcripts are deposited in a public repository or must have clear agreements in place for how long the transcripts and/or tapes are to be closed to researchers, and must warn interviewees about the risk of disclosures. Interviews that the researcher, the interviewee, and the repository agree to close until the interviewee's death, for example, may be subject to disclosure to law enforcement through a subpoena, and the interviewee needs to be told this.[30] If an IRB decides that the possibility of harm to the interviewee (or to the historian, when interviewing people about undisclosed past crimes) is too great, the project can be halted before it begins. And there is no appeal.[31]

One of the gray areas left unaddressed by the *Belmont Report* and the Common Rule, focused as they were on biomedical and behavioral research in the 1970s and 1980s, is the problem of "third party" harms. These are harms to individuals who are not directly involved in the research, but whose interests are drawn into it by the research subject.[32] The now classic example is research involving genetic analysis. Genetic information about a single individual may reveal information about uniquely identifiable relatives, especially parents, siblings, and children, without their consent.[33] Similarly, in behavioral research involving intimate relationships, the research subject may disclose private information about family, friends, or others without their consent. If all of this information is kept confidential and anonymized, then risks are minimized and no one worries overmuch. While IRBs have no formal responsibility for evaluating harms to third parties, ethicists urge researchers to be attentive to such possibilities and to deal with them explicitly in their research designs. Some urge OHRP to enlarge the scope of IRB review to include third-party harms, possibly even to require third-party consent in some cases.[34]

Oral historians, of course, interview people who talk frankly about those they have known. Telling stories about personal relationships, work conflicts, or trouble with authorities brings up the names and actions of others. The historical record would become strangely muddied if interviewers had to censor interviewees' memories and opinions out of fear that someone could be damaged by public access to oral history transcripts and audio files. As far as I have been able to determine, no one has yet suggested that this would be an appropriately ethical thing to do.[35] It has been left, instead, for those who have felt they have been harmed to seek other options: they can sue. I cover that path in chapter 3. Worry over third-party harms has certainly affected archivists, however, who regularly deal with letters, diaries, and other documents in which the author reveals intimate details about others. I cover their concerns in chapter 4.

In research involving children and the elderly, both considered vulnerable populations, promising to keep information about third parties confidential runs up against state laws that require doctors, teachers, caretakers, and other responsible parties to report evidence of possible abuse and neglect to the proper authorities. Even though few state laws list "researchers" among those required to report abuse or neglect, some IRBs have determined that they must, and so researchers need to reveal that possibility in consent documents.[36] In other words, the "third parties" at risk here are those who abuse children and the elderly, and they get no sympathy under these laws. Ironically, then, when parents are asked to consent to include their children in research studies that might reveal abuse or neglect, they need to be told of the risks to themselves, and to other adults, if a child discloses—even without realizing it—harms done to them by those adults. The risk to the adults, of course, is being reported for child abuse. Oral historians rarely interview children, but they often seek the memories of the elderly, some of whom might be dependent upon caretakers for their physical needs. Technically, then, on the off chance that an oral historian might suspect, witness, or be told about abuse or neglect during an interview, a very cautious historian should warn her elderly subjects that she may report it.[37]

A similar gray area persists in the secondary research use of data collected for another purpose. What distinguishes the questions about "secondary" data use from those surrounding interviewing people is the fact that using "secondary" data does not involve the researcher interacting with living human beings, only with information about them. Information about people has been piling up in all sorts of government, corporate, administrative, and research databases and archives for decades. Some databases contain information linked to identifiable individuals, such as those held by the Internal Revenue Service; others contain completely anonymized grouped data.[38] The vast majority of databases are restricted because they are protected by privacy laws, were compiled from researchers' own collection of data, or have economic value to marketing, advertising, and other corporate interests. IRBs must scrutinize researchers who wish to use restricted data sets to ensure that they have proper security measures at hand to keep the data protected and that all staff people are trained to know that they are not to disclose any of the data to others. In cases where identified data were collected for one research project and then a researcher wishes to use them for another research project, or to share them with other researchers, IRBs wrestle with the ethical problem of consent. The subjects only consented to the first project, so shouldn't they be contacted for consent for "secondary" use? IRBs are also concerned about the probability that individuals may be identifiable when data sets involving the same geographical area are merged with each other.[39] In a small town, an anonymized community survey that includes responses about income ranges, gender, age and family size when merged with public property tax records, for instance, could reveal how rich old childless Mrs. Vlasdavostoc answered a set of survey questions about (for instance) school funding.

Technically, data collections identified as "publicly available"—because they are literally released for public use by some organization or agency, such as the U.S. Census Bureau's aggregated data sets—raise no ethical issues and no IRB review is required to use them.[40] The underlying assumption is that the agency, group, or individual who released the data to the public has followed the various laws protecting identifiable private information gathered by the government or by businesses, and so the data have, in a sense, been pre-screened for appropriate use.[41] By default, the Common Rule treats documents ("data") in government funded (state, city, county) archives as "publicly available" since the public has access to them (although some collections may be closed for a period of time), and has left any privacy issues about deposited materials ("data") to archivists and manuscript curators. Materials deposited in privately funded archives may be completely open (most private universities), completely closed (corporate archives, such as those held by Motorola), or have some open and some restricted areas (the collections of the American Medical Association, for example), and their archivists manage who sees what, when, and how.[42] I deal with archivists' responsibilities and decision-making authorities in chapter 4; here it is enough to stress that IRBs have not yet forayed much into demanding prior review of archive-based projects.

But they could. As long as a collection that includes identifiable personal information about individuals might include living people, then using it might require IRB staff scrutiny to determine if the historian's project is not research (not seeking generalizable knowledge), exempt from review, or suitable for expedited review, even in publicly available data.[43] At the University of Chicago, for instance, "research involving the analysis of publicly available data containing private identifiable information . . . is exempt from IRB review. University and SBS [Social and Behavioral Sciences] IRB policies identify the IRB as the unit responsible for determining exemption from IRB review. As such, research involving these analyses requires the submission of a protocol for consideration by the IRB."[44] Taken to a literal extreme, the University of Chicago's policy could be applied to a set of letters from the 1940s deposited in the Chicago Historical Society's archive if they contain personal details about identifiable people and if an IRB decided that it is responsible for any living people who might be mentioned in them. In an even more explicit statement, Winthrop University's policies declare that "any study using archival documents that are publicly available, such as library archives or legal cases [is exempt]." Who decides the project is exempt? The IRB office, by viewing an initial proposal.[45] To what extent historians at either institution are actually complying with these IRB expectations is an open question.

Human subjects' regulations apply only to the living, but their implications spill over onto historical research more generally. First, as historians we should be concerned that any significant research method, such as oral history, has been made unnecessarily difficult to accomplish, for fear that students and researchers will forgo interviewing witnesses to history when faced with the alien world of IRBs and regulatory language. The idealistic solution to that problem may be to educate ourselves better about IRB requirements, and to educate IRBs about history as a research discipline. Second, the expansion of IRB interest in requiring pre-review of studies involving archival materials that may contain personal information about identifiable living people, even if most of the identifiable people are dead, will require many more historians to submit their proposals to IRB offices, with the attendant frustrations over the paperwork and time required before research is allowed to begin. The possibility that an IRB office could require a historian to document the death of individuals named in archival documents before being allowed to use names, moreover, would place considerable burdens on researchers.

The third reason is far more important than simply the extra work placed on historians, work that other researchers and IRBs may find quite appropriate. The saga of human subjects regulations tellingly illustrates the power of regulatory language and the tendency, as several observers have noted, for bureaucratic control to creep from its originating purpose (biomedical and behavioral research) into neighboring areas (oral history) as that regulatory language is re-interpreted and applied ever more rigorously.[46] When HIPAA extended the umbrella of privacy over the medical records of the dead in perpetuity, a policy that lasted from 2003 to 2013, precedent was set for protecting other kinds of sensitive information about the deceased. The range of privacy laws passed by Congress in the last fifty years suggests what the political public has considered particularly important to protect. The possibility that privacy protections could be explicitly granted to decedents' records in these areas may well seem extremely remote—perhaps as remote as IRB scrutiny of oral history seemed just a few decades ago.

Privacy and Federal Laws: Information Acts

Since the 1960s, the U.S. federal government has passed a hodgepodge of laws that apply to information about citizens and residents. These information laws nearly always mention privacy in one context or another, and so become of interest to the historian probing means of access to, and disclosure of, information about other human beings, living or dead. Table 2 summarizes the major laws, and invites a short detour into the legislative and regulatory process. When Congress passes legislation and the President signs it, it becomes a public law and gets a public law number. Every public law is a statute, and so enters the collection called the United States Statutes at Large. Every six years, the Office of the Law Revision Counsel of the U.S. House of Representatives publishes the revised United States Code (USC), which reflects all of the changes made in U.S. law by the public laws passed. A new public law may fit nicely into one area of the USC or may actually affect different sections of the USC, with the law as written dispersed among the Titles to which it applies. If this were not complicated enough, much law is never passed by Congress. Various agencies in the Executive Branch have the authority to create the regulations needed to actually make statutory laws work, and these are created by a rule-making process in which drafts of regulations are published in the *Federal Register* with an open period of time for citizens to comment upon them. Then the agency reviews the comments, makes changes to the draft regulations that its staff thinks are appropriate, and publishes the final rule in the *Federal Register*, setting a date for it to go into effect. The rules then become part of the Code of Federal Regulations (CFR). Agencies can amend their regulations by going through the same process as often as they wish.

The National Biomedical Research Fellowship, Traineeship and Training Act, for example, better known as the National Research Act, was passed in 1974 as Public Law 93–348 or 88 Statute 348. United States Code Title 42–The Public Health and Welfare, chapter 6A–Public Health Service, subchapter III–National Research Institutes, part H, section 289 gives the Secretary of Health and Human Services the authority to write and enforce regulations concerning human experimentation and informed consent. The resulting regulatory laws establishing IRBs then appeared in Title 45 of the Code of Federal Regulations (CFR), which includes all of the regulations that apply to public health. All of the other laws that affect how certain kinds of information are controlled have similarly convoluted histories, and appear more or less coherently in the text of the United States Code and the Code of Federal Regulations.

The cluster of privacy laws manage information from two directions: making it open and keeping it closed. In all cases, the laws—and their associated federal regulations—aim to prevent harm to individuals, while at the same time allowing the public to scrutinize government actions and the government to support law enforcement and protect national security. The passage of the first Freedom of Information Act (FOIA) in 1966, for example, was a milestone in the opening up of the federal government's records to citizens' inspection. FOIA requires the Executive branch of the government to produce documents requested by anyone, for whatever reason they wish to see them, unless they are protected by certain restrictions, notably concerns over individual privacy and national security. The basic assumption is that documents will be released, so the government has to have compelling reasons to refuse. Every agency's section in the Code of Federal Regulations includes information on how FOIA requests may be made. FOIA has been a major asset to historians of twentieth-century American political and social history, at least for those who have the patience to use it (see the further discussion later).[47]

Table 2 Major Federal Legislation: Control over Information

Year	Legislation or Regulatory Code	Code	Regulations
1934	*National Archives created*	Public Law 73–432	
1954	*Bureau of the Census created as ongoing agency within the Department of Commerce*	13 USC	
1966	*Freedom of Information Act (FOIA), Exemption 6* Denies release of medical records or "similar" files that "would constitute clearly unwarranted invasion of privacy."	5 USC § 552b	Every agency's CFR
1968	*National Archives and Records Administration made an independent agency* The regulations codify retention and transfer policies for the records of federal agencies and clarify the power of the Archivist's office to determine which records have historical value.	44 USC 21	36 CFR Chapter XII
1970	*Currency and Foreign Transactions Reporting Act (Bank Secrecy Act)*	31 USC § 5311–5314e	31 CFR Chapter X
1970	*Fair Credit Reporting Act* Enables regulations to cover the responsibilities of credit reporting agencies to consumers. Amended several times.	15 USC § 1681	Various financial agencies and commissions
1974	*Family Educational Rights and Privacy Act of 1974 (FERPA)* Outlines parental and individual rights to access "educational records" and restrictions on disclosures to third parties.	20 USC § 1232g	34 CFR Part 99
1974	*FOIA, Exemption 7(C)* Extends denial of the release of records gathered and held for law enforcement purposes if the release "could reasonably be expected to constitute an unwarranted invasion of personal privacy."	5 USC § 552b	Every agency's CFR
1974	*National Biomedical Research Fellowship, Traineeship and Training Act (National Research Act)* (See Federal Policy for the Protection of Human Subjects, 1981.)	Public Law 93–348	
1975	*Privacy Act* Addresses post-Watergate concern for government collection of information on individuals; requires the government to disclose information held about a person to that person.	5 USC § 552a	Every agency's CFR

Year	Legislation or Regulatory Code	Code	Regulations
1978	*Right to Financial Privacy Act* Outlines access rights that government agencies have to individuals' financial information and responsibilities of financial institutions to notify customers of disclosures.	12 USC Chapter 35 [§§ 3401–3422]	28 CFR 47; 32 CFR Part 275 and various
1980	*Privacy Protection Act* Explicitly forbids the government from seizing materials from a person who has them in order to prevent publication, unless the person is committing a crime, or intending to publish classified data or child pornography.	42 USC 21A	
1981	*Federal Policy for the Protection of Human Subjects* Establishes the requirements for ensuring the consent and safety of individuals who participate in research as "subjects," including the confidentiality of research information collected. Amended multiple times.	(See National Research Act.)	45 CFR Part 46
1984	*Cable Communications Policy Act* Deregulated the cable industry. Includes consumer privacy protections.	47 USC 551	47 CFR 76
1986	*Electronic Communications Privacy Act* Deals with the privacy of wireless voice and electronic communications for law enforcement, business, and other purposes. Amended several times.	18 USC §§ 2510–2521, 2701–2711	
1988	*Video Privacy Protection Act* Makes it a crime for businesses to release records about individuals' rental of video [DVD] recordings.	18 USC § 2701–2712	
1991	*Telephone Consumer Protection Act* While this act mainly concerns automatically dialed calls, it includes protections for consumer privacy against unwanted telephone calls.	47 USC § 227	47 CFR 64
1994	*Driver's Privacy Protection Act* Made it a crime to disclose personal information held by state departments of motor vehicles to people without the driver's consent, except for law enforcement and other purposes.	18 USC Chapter 123	

(continued)

Year	Legislation or Regulatory Code	Code	Regulations
1994	*Communications Assistance for Law Enforcement* Requires telephone and other communications companies to make it possible for law enforcement to perform electronic surveillance in real time.	47 USC §1001–1010	
1996	*Telecommunications Act* This overhaul of telecommunications law basically deregulated broadcasting and covered broadband services for the internet. It includes privacy provisions for consumers.	Public Law 104–104	
1996	*Electronic Freedom of Information Act Amendments* Expands citizens' access to government records held in electronic form.	5 USC § 552	
1996	*Health Insurance Portability and Accountability Act (HIPAA)* Requires the government to establish regulations to govern the development and use of electronic health records.	42 USC § 1320d	
1998	*Children's Online Privacy Protection Act* Requires websites to include details about the collection of information from children under thirteen in their privacy policies, covers parental consent requirements, and restricts marketing to children.	15 USC § 6501–6506	
1998	*Computer Matching and Privacy Protection Act* Adds restrictions to the Privacy Act on how data about individuals in different government record systems can be matched and linked through computer software.	5 USC § 551a(o)	Every agency's CFR
1999	*Financial Services Modernization Act* Elaborates on the privacy protections that financial institutions regulated by the Federal Trade Commission, the National Credit Union Administration, the Securities and Exchange Commission, and state insurance companies owe to customers.	15 USC §§ 6801–6810	CFR of agencies listed
2001	*Standards for Privacy of Individually Identifiable Health Information (mandated by HIPAA) published* Went into effect in April 2003; modified 2013.		45 CFR Parts 160 and 164

Year	Legislation or Regulatory Code	Code	Regulations
2001	*Uniting and Strengthening America by Providing Appropriate Tools Required to Intercept and Obstruct Terrorism Act (USA Patriot Act)* Passed following 9/11, this legislation amended more than twelve sections of the United States Code in order to increase law enforcement's power and flexibility to gather information about citizens and residents in pursuit of possible terrorist groups and acts.	Public Law 107–156	49 CFR Parts 107, 171, 176, 177, 383, 384; 42 CFR Part 73
2008	*Genetic Nondiscrimination Act*	Public Law 110–233	29 CFR Part 1635
2013	*Heath Information Technology for Economic and Clinical Health Regulations (HITECH Act)* Part of the American Recovery and Reinvestment Act (2009). It applies specifically to health care information technology designed to further encourage the development of electronic medical records.	Public Law 111–115	42 CFR Parts 412, 413, 422, and 495; 45 CFR Subtitle A Subchapter D

Note USC = United States Code; CFR = Code of Federal Regulations. For the most current versions of these codes, see the Government Printing Office website, GPO Access, at http://www.gpoaccess.gov/uscode/index.html for statutes (legislative branch) and http://www.gpoaccess.gov/cfr/index.html for regulations (executive branch).

The Privacy Act of 1975, in contrast, put restrictions on what the government can do with the information it collects on individuals in order to protect their privacy. At its core, the act prohibits the government from melding data bases in which individuals are identified across agencies, or transferring information collected by one agency (e.g. Internal Revenue Service) for one purpose, to another agency (e.g. Department of Justice) for another purpose, without the consent of the individuals involved. Having bits of information scattered in different government records, in other words, is not seen as a privacy problem until they are linked together with the potential to form complex individual profiles.[48] The Act's proponents recognized that people also needed an explicit right to learn exactly what information the government has on them in any one of its myriad agencies, to correct that information if it is wrong, and to seek redress if incorrect information is not removed from an agency's records.[49] Along the way, of course, the Privacy Act specifies all sorts of exemptions and departures from these overarching principles of protecting privacy and freedom to know what the government knows about a single person. The exemptions for data melding, for example, include all of the government's needs to conduct "routine" business, as well as allowable searches for law enforcement and fraud detection purposes.[50] Information about an individual in a government file, in turn, may be exempt from disclosure to that individual in the interests of law enforcement or national security. So, every agency's Code of Federal Regulations has a section on the application of the Privacy Act to its records.[51]

The rest of the laws in table 2 apply the same balancing acts of disclosure versus privacy to the specific kinds of records they cover, from educational records in public schools (Family Educational Rights and Privacy Act) to video rentals (Video Privacy Protection Act) and telephone accounts (Telephone Consumer Protection Act).[52] While some of these acts focus on privacy as a central issue, others (such as the Telecommunications Act) had other targets and included privacy protections simply as part of the overhaul of the systems that the legislation addressed. Two of the acts, in contrast, explicitly counter privacy claims with the needs of law enforcement (Communications Assistance for Law Enforcement, 1994) and national security (Uniting and Strengthening America by Providing Appropriate Tools Required to Intercept and Obstruct Terrorism Act, a.k.a. U.S. Patriot Act, 2001), although neither of these laws allow information to be divulged to those outside specific categories of government agencies.

How these laws are enforced provides further insight into the tools that legislators and regulators can deploy to ensure compliance. If the legislation applies to the actions of government (FOIA, Privacy Act, FERPA, Common Rule, HIPAA), then the enforcement stick is the power to penalize the wayward agency or to withhold federal funding from the offending institution, such as a public school or a research university.[53] At least for FERPA and HIPAA, moreover, the courts have firmly ruled that individuals who believe they were harmed by the inappropriate disclosure of information under these regulations simply do not have a "private right of action."[54] That is, they cannot sue the individuals who transgressed, or their parent institutions, directly. Employees who mess up can, of course, be fired, but that punishment does not pay damages to an allegedly wronged individual. In all of the laws that apply to the private sector, in contrast, such as financial institutions, telecommunications companies, and video rental outlets, clauses explicitly give individuals the right to seek civil penalties from the institution or company whose employees transgressed.[55] These technicalities matter when it comes to the real (or imagined) threats that historians could face if they were to disclose the names of living people found in records covered by one or more of these laws (see more on being sued in chapter 3).

So, then, what do these laws mean for historians?

First, these laws emphasize the late twentieth-century turn towards statutory and regulatory law as a way to deal with social relations, a trend that shows no signs of abating as concerns about privacy, data and the internet continue to proliferate among consumers and citizens.[56]

Second, they do, and will, affect the kinds of documents that archives can be expected to acquire and policies about restrictions on access during the lifetimes of named individuals. None of these laws or regulations applies to the records of the dead, except for those contained in the modified HIPAA Privacy Rule (2002) of 2013, which currently protects decadents' records for fifty years after the date of death.[57] Many a time I have heard both historians and educators say that FERPA, for instance, applies to all student records for all time. Yet, FERPA does *not* apply to the records of the deceased.[58] It also does not apply to private schools that do not receive federal funding. State laws, of course, may cover the records of deceased students, and archivists and historians certainly should find out.[59] Nevertheless, federal privacy laws have less range than many may assume. Anecdotal evidence suggests, for instance, that what people believe about the scope of privacy laws has a chilling effect on their willingness to save and to donate old family or business papers to an archive.[60]

Finally, except for the Common Rule and the HIPAA Privacy Rule, none of these acts or regulations specifically considers access to records for research purposes. Even when the law mentions the accumulation of de-identified information into databases, their allowable disclosures refer to use for "supervisory functions" for the "supervisory agency"—as in the Right to Financial Privacy Act —or for similar oversight requirements by the agency in charge of regulations, from the Department of Justice to the Federal Communications Commission.[61] The supervisory agency—the Federal Reserve or the Department of Education, for instance—can then release very large anonymized data sets for public use and can provide access to restricted data sets "for qualified researchers" if it so chooses.[62] Such aggregate data serves the needs of social science researchers in a wide variety of fields, and some of the largest (such as those of the National Center for Health Statistics, National Center for Education Statistics, and the collections of the Inter-University Consortium for Political and Social Research) have been "pre-approved" by major university IRBs for use without any submission of research protocols to an IRB office.[63] Current and archived data from these collections are priceless for social historians of the United States in the twentieth and twenty-first centuries, as long as the researcher is interested in large populations rather than in individuals or lowly populated areas. For everything else, there's FOIA.

FOIA: Getting Information Out of Government

The federal Freedom of Information Act deserves attention in a study of history
and privacy because the government can refuse to release information due to
the possibly damaging effect on (living) individuals' privacy. Every state has its
own laws that apply to state and local government documents, moreover, and
these can vary widely, so historians seeking their records must delve into state
statutes.[64] There are a number of guides on how to submit FOIA requests and,
as noted earlier, every federal agency is required to make its own process available to citizens.[65] Here the focus is on how the law has worked—or not—for
historians submitting FOIA requests and, in key instances, for others asking for
documents that historians might want, from the perspective of legal decisions
concerning privacy.

While a great many of the documents produced by federal agencies are regularly deposited in the National Archives and Records Administration (NARA)
and must become available no more than thirty years after their creation, by
no means are all released into the public domain. Huge bundles are destroyed:
documents of no historical interest (according to NARA experts), those for
routine business, personnel files for ex-employees, and the like. Agencies may
also keep some old documents (or other media) in-house, notably those at the
Department of Justice and the Department of Defense, for instance. It is these
documents, as well as more recent ones, that are subject to FOIA requests.
Agencies may refuse a request, for which it must give good reasons, based on
formal exemptions written into the enabling law.

Information may be withheld "in the interest of national defense or foreign
policy," or if it is protected by explicit clauses in other statutes. Trade secrets stay
secret.[66] "Personnel and medical files and similar files the disclosure of which
would clearly constitute unwarranted invasion of privacy" (Exemption 6) are
restricted. Law enforcement records are open, unless "enforcement proceedings"
are ongoing, or they might "deprive a person of a right to a fair trial," or could
"disclose the identity of a confidential source," or release information provided
by a confidential source that could identify the source, or reveal law enforcement techniques that could "risk circumvention of the law," or "endanger the
life or physical safety of any individual." Included in this list, too, is the fairly
open-ended exemption for law enforcement material that "could reasonably be
expected to constitute an unwarranted invasion of privacy" (Exemption 7[C]).[67]
Names and identifying details might be redacted in such cases, depending on the
record, so that the citizen receives some information; other records will simply

be denied. To overcome the privacy restrictions, the applicant has to demonstrate that a possible invasion of privacy is warranted.

The courts get involved when citizens believe that the federal agency is stonewalling and not releasing records subject to FOIA. In 1989, for example, the Supreme Court squashed requests by Peter Irons (whose *Justice at War: The Story of the Japanese-American Internment Cases* had partly depended upon FOIA-released sources) for FBI documents on Smith Act prosecutions during the McCarthy era. The court upheld the FBI's assertion that information in the documents came from confidential informants.[68] Even if the informants were likely dead, the court agreed with the FBI that it was vital to prevent any chilling effect that releasing such information might have on living informants, who could fear future reprisals against family members if it were known that their identities could be released after their deaths.

In 2003, the United States Court of Appeals for the District of Columbia refused Ellen Schrecker's FOIA request to reveal the names of individuals redacted from FBI files she had requested and—after many years—had been given. She was also researching FBI investigations during the McCarthy era. If the individuals were deceased, their names could be released. The problem was proving their deaths. The FBI looked up names in *Who Was Who*, noted birth dates found in the files, and used "other readily available information" to check for decedents. If only dates of birth were found, the FBI was prepared to release names if one hundred years had passed since an individual's birth date. Schrecker wanted the FBI to search more broadly in their files for dates of birth and/or Social Security numbers for those individuals and to check Social Security numbers against the Social Security Death Index. The FBI responded that doing so would be an undue burden on its resources. The FBI asserted that it performed reasonable searches in the files already found—some twenty-four thousand pages of material—and had checked easily accessible sources. Any individual whose name and date of death was not found in one of these sources, or had no birth date given, was assumed to be alive. Despite presenting a statistical argument on the relatively small chances that adults living in the 1950s are alive in the early twenty-first century, the plaintiff made no headway with her claim that the FBI was being unreasonable.[69]

Schrecker v. Department of Justice (2003) confirmed, again, that the FOIA privacy exemption applies only to living people. The restriction on releasing information that could identify a confidential source even if that source were dead is based not on concerns for the source's privacy, but on the needs of law enforcement to encourage people to come forward and provide information about crimes. In 2004, nevertheless, the Supreme Court affirmed that the FOIA privacy exemption could be applied to protect the privacy of living relatives of a deceased person, at least in the special circumstances of death scene photographs. When Vincent Foster, one of President Clinton's deputy counsels, died in 1993, several agencies investigated and determined that he had committed suicide. A lawyer, Allan Favish, claiming that he was not satisfied with these conclusions, submitted a FOIA request for photographs that had been taken at the scene of Foster's death. Foster's family objected strongly, and asserted that they were protected by the FOIA exemption (7[C]) for privacy. The case wound its way through the District Court for the District of Columbia, the Court of Appeals for the District of Columbia, and finally reached the Supreme Court in 2003. Justice Kennedy, writing for the unanimous court in 2004, decided that the death scene photographs should not be released. He explained that Congress had written the FOIA exemption quite carefully, and agreed with the district court that "the statute's protection 'extends to the memory of the deceased held by those tied closely to the deceased by blood or love.'"[70] Foster's privacy was not the issue—he had no rights, as he was dead. But Foster's relatives, describing the media onslaught that had followed them and the pain caused by the release of one leaked photo onto the internet, made a case that "their own right and interest to personal privacy" should prevail.[71]

The court recognized that the need for public scrutiny of law enforcement is an important value, and one not to be denied lightly. To support the protection of death photos, Justice Kennedy fell back on the common law tradition of "the right of family members to direct and control disposition of the body of the deceased."[72] He also evoked the importance of respect for dead bodies embedded in burial rituals. Favish argued forcefully that disclosing the pictures was in the public interest because he suspected the government had behaved improperly in its investigations, and might even be covering up Foster's murder. Justice Kennedy was not convinced, since so much else about the investigations had been released, and Favish had no evidence to support his suspicions. Favish, Kennedy stated, had to show sufficient reason that releasing the materials was in the public interest, and he did not reach that standard. There was no way for Favish to see the photographs by himself in order to make a personal judgment, moreover, because under FOIA, "once there is disclosure, the information belongs to the general public."[73]

The Favish decision concerns FOIA advocates for three reasons that historians should heartily endorse. First, it imposes a "sufficient reason" standard that applicants might be asked to meet if an agency refuses documents under the exemption for individual privacy. The information disclosed must be shown to have "significant public interest." If an applicant claims that the material will help to uncover government wrongdoing, then he or she must have prior evidence to support that claim. The FOIA text itself makes it clear that citizens do not have to have a reason to ask the government for information, but they do if the government refuses to release it, and now suspicion of a cover-up may not be a good enough reason.[74] Second, the more the right to privacy is invoked as a reason to withhold documents, the stronger it gets as a foundation for a government agency to refuse requests. Privacy claims could thus be a way to enhance government secrecy, and so to circumvent FOIA's intent.[75] Not many people have the resources to challenge the government over a refused FOIA request. Third, it clearly confirmed that the privacy interests covered by FOIA extended from the individual to the deceased individual's family when that information was especially graphic and hence potentially very painful to survivors.[76] Whether courts will be inclined to broaden the meaning of the privacy interest that "extends to the memory of the deceased held by those tied closely to the deceased by blood or love" beyond gruesome photographs to other sorts of unpleasant material remains an open question to be decided only after further FOIA refusals and lawsuits. Let us hope the courts resist the temptation to further protect "the memory of the deceased."

Challenging the Common Rule and the HIPAA Privacy Rule in Court: A Case Study

To date, no researcher has challenged the authority of the federal government to regulate social science research in a suit that has reached a state or federal court of appeals. If researchers have done so in local courts, either the cases have been dismissed, settled out of court, or went to trial with neither party appealing the court's decision.[77] Similarly, no researcher has yet challenged HIPAA's restrictions on access to medical records for research purposes. A handful of cases at the state level have, however, pitted clauses in states' open records laws against HIPAA regulations when state agencies have refused to release medically related information by claiming that the HIPAA Privacy Rule prevents them from doing so. One case in particular is important for archivists, genealogists, and historians, as it not only outlines what it took to challenge a federal regulation in court, but also illustrates arguments for and against releasing the names of some of those who died at a mental institution.[78]

    In April of 2007, the Adams County Historical Society in Nebraska formally wrote to Marj Colburn, the operating officer of the Hastings Regional Center, a substance abuse treatment center for juvenile boys, and asked her to release the names of the approximately one thousand individuals who were buried in the center's cemetery between 1888 and the 1950s. The graves there were marked only with numbers. The campus of the Hastings Regional Center had variously been the Hospital for the Incurably Insane, the Nebraska State Hospital, the Ingleside Hospital for the Insane, and the Hastings State Hospital, since its founding in 1887. It was well-known that the cemetery there contained the bodies of those who had died at the hospital and whose relatives and friends were unable to claim them for private burials. The Adams County Historical Society wanted to make the names available to genealogists, and to provide a way to link individuals to their graves. The request was pushed up to the Nebraska Department of Health and Human Services (DHHS) and, in due course, Nancy Kinyoun, the DHHS information officer, refused to release the names due to state laws against the release of medical records and, more importantly, on the basis of HIPAA's 2003 Privacy Rule, with its expansive definition of protected health information. According to the HIPAA Privacy Rule, such information includes all identifiable information about the "past, present, or future physical or mental health or condition of an individual [or] the provision of health care to an individual."[79] This means that even the information that a person is, was, or will be a patient of a health care provider (e.g. doctor, hospital) is protected.

The Adams County Historical Society (ACHS) protested and eventually took the matter to the District Court of Adams County, with the significant help of an attorney, Thomas Burke, who was willing to do the heavy lifting for the historical society.[80] According to Nebraska state law, burial records are public records. Death certificates are public records, too, and they give cause of death as well as location of death, which was more than the ACHS was asking for. The ACHS only wanted to know where dead people were buried; no one asked for any details about how they died. The ACHS also stressed the historical context: the burials appeared to span the decades between the late 1880s and the 1950s, a time in which "domestic trouble" and "religious excitement" brought individuals to the institution. No one could say that particular patients had been, in fact, mentally ill in a current diagnostic sense. The ACHS further pointed out that records for other cemeteries where the center's patients had been buried included the fact that they had died at the center—and that those were clearly recognized as public records. In February of 2008, nevertheless, the district court agreed with Nancy Kinyoun that releasing names of the dead buried in the cemetery was the same as releasing the fact that they had been patients in what was then a mental institution, that HIPAA protected that information after death, and that HIPAA trumped state public records laws.[81]

The ACHS appealed to the Nebraska Supreme Court, petitioning the court to require the release of the information because the district court had erred in its interpretation of Nebraska's public records statutes and in its application of the HIPAA Privacy Rule. A clause in the Privacy Rule explicitly allows for the disclosure of information where "the disclosure is required by law" including state laws; cases from other states had already upheld the position that HIPAA could not prevent disclosures under established public record acts.[82] Nancy Kinyoun (i.e. the DHHS) counter-appealed, claiming, among other things, that the ACHS had no business appealing because they were not relatives or personal representatives of the dead people. Any person who wanted to know if she had a relative buried at the center would know from the death certificate where her relative died. She could then hire an attorney and file an appropriate legal request with the DHHS to release the location of the relative's grave, and hence to confirm that the relative had been a patient there. The ACHS explained flaws inherent in this method: it is not always possible to find death certificates, as these were not always filed in the period under discussion; significant variations in spelling of names, moreover, even when death certificates were filed, sometimes made it impossible to find them. The ACHS also noted that it placed an undue burden on a person to have to hire an attorney simply to get a burial record. Finally, the ACHS pointed out that Nebraska has a serios of statutes dedicated to maintaining burial records, especially historical burial records, and for preserving cemeteries and the graves of veterans, homesteaders, and others. "All human burials are accorded equal treatment and respect for human dignity without reference to ethnic origins, cultural backgrounds, or religious affiliations" the ACHS brief quoted from Nebraska law.[83] Why exclude patients of mental institutions from this respect?

In May 2009, the Nebraska Supreme Court decided that the ACHS's request for names amounted to a request for burial records which were, in fact, public records, and so not covered by any of Nebraska's privacy laws for medical data; they were not covered by HIPAA because burial records fell among those whose "disclosure is required by law" in the Privacy Rule.[84] That was the legal basis for the decision. But both sides had also raised public policy arguments that evoked some of the ethical themes embedded in the Common Rule. During oral arguments at both the district court's and the Supreme Court's hearings, for example, the attorney for the DHHS claimed that disclosure of the identities of those buried at the Hastings Regional Center would cause harm: harm to principles of medical privacy and harm to the relatives of those revealed to have been patients at a mental institution. The DHHS sought to minimize stigma by protecting those potentially stigmatized after death; making those names public should be a decision for the next of kin, not for society in general, nor for the ACHS in particular.[85] The attorney for the ACHS countered that the records sought were for deaths more than fifty years ago, and some were for people whose immediate families had abandoned them. To keep those names secret simply perpetuated the stigma of mental illness. The ACHS also pointed out that the names of people who had died in mental hospitals, and whose graves were marked, had been public for decades in various historical societies' collections and that no complaints had come to anyone's attention. The harms that DHHS claimed were clearly speculative, as demonstrated by the fact that the DHHS submitted no evidence to support them. In contrast, genealogists' curiosity about their relatives, and their need to understand their ancestors' lives, was well-known.

It took a little over two years for the ACHS to get the names and grave locations for those buried in the Hastings Regional Center Cemetery, from the initial formal request to the Hastings Regional Center to the DHHS's final compliance with the Nebraska Supreme Court ruling.[86] This was a remarkably fast turnaround, and it speaks to the persistence of the ACHS's team in filing their claims and appeals. This case, like the FOIA rulings discussed earlier, illustrates the challenges involved in contesting the power of federal (and state) regulations to control access to information. It further illustrates the way that regulations designed to protect the privacy of current information about living individuals can creep over records largely of interest to historians and others concerned about the past, including records about the dead. In this case, pushing back against the use of HIPAA to prevent the release of information about where hospital patients' bodies were buried means that one perceived public good (burial records are public records) had more weight than another perceived public good (all knowledge that a person had been a hospital patient is protected health information). What was really at stake, perhaps, was the fact that the hospital had been created and maintained for the mentally ill. Would the Nebraska DHHS have gone to the mat for patients buried at a tuberculosis sanatorium? Or an old-folks home? Or a county home that had once been a poor farm?

Conclusion: Historians and Regulatory Law

At the moment, historians who work at universities subject to IRB regulations and who plan to interview people for their research need to contact whatever campus offices handle IRB work and inquire about local requirements for oral history projects. They risk having their research halted if they proceed with interviews without properly approved informed consent documents or, at least, affirmation that their projects are exempt from review.[87] No one, as far as I have determined, has faced similar IRB problems with research in archives where the collections consulted might contain information on living people, but the potential for them exists. The complex network of access and privacy laws for information held by the federal government (and their parallels in state government laws and regulations) and regulated industries further shape the universe of resources that historians can access (without restrictions) for their work.

The trajectory of regulatory law and IRB bureaucracy is disquieting but unsurprising. In part, that is the nature of bureaucracies inspired by the penalties for failing to enforce their regulations rigorously enough. When the government can shut down all federally funded research at an institution for apparent violations of the Common Rule, or an IRB can deny a doctoral degree to a student who did not properly submit a research protocol and consent forms to it in advance of gathering data, it rarely matters that such actions are rare. The suspicion, too, that researchers might try to get away with something unethical if not monitored—and the strong sense that because researchers have to be monitored, they must be untrustworthy—maintains the current system. Federally mandated oversight has embraced history, and historians, both through laws to protect living human beings as research subjects and laws to protect the medical privacy of the dead. Those laws, in turn, support a set of beliefs about how to behave ethically that go far beyond the literal scope of the laws themselves.

Even if IRBs continue to turn a blind eye to the possible harms that could come to a living person if personal details about his life are exposed by a historian without that person's consent, the living person has options in law. Historians, like journalists and other authors, can be sued for invasion of privacy, specifically for "public disclosure of private facts." At this point, it should be clear that research ethicists would chorus that it is better to prevent harm than to punish those who cause harm. Once the "private facts" have been released to the world, the researcher/journalist may have caused real suffering (even if minor) and has not shown proper respect for persons by seeking informed consent. In the next chapter, I explore the reasons that courts have resisted punishing authors for the "public disclosure of private facts" even as they acknowledge that authors have been rude, insensitive, and careless.

Chapter 3

Historians, the First Amendment, and Invasion of Privacy

Journalists' freedom to interview people and to publish with exceedingly few restrictions provides an important contrast with researchers' need to comply with Institutional Review Board (IRB) requirements when interviewing research subjects in the process of gathering data for "generalizable knowledge." The bulwark that makes it very hard to actually punish journalists for publishing true information (truth being a key defense against libel and defamation) is, of course, the First Amendment to the Constitution of the United States, discussed in the first part of this chapter. Once they publish, journalists and their publishers then run the risk of being sued for libel, defamation of character, and/or invasion of privacy if a person believes that she has suffered one of these injuries. These injustices are covered in the law of torts, dealt with in the second part of the chapter. The contrast with chapter 2 may be succinctly summarized: in research, exposing people to the risk of harms is to be minimized; in journalism, committing actual harms is to be punished.[1]

Historians, like academics in all sorts of fields, conduct research and write for open publication (as opposed to writing confidential reports or trade documents). As explored in chapter 2, historians might cause harm at two points in their work: first, when interviewing people about especially sensitive or risky topics or gaining access to confidential or private information (data) without appropriate safeguards in place to protect it; and second, if they disclose that sensitive, confidential, or private information in ways that allow living individuals to be identified to those who should not learn such things about them, such as the world at large. These are obviously interconnected. If a historian's research protocol and consent form for interviews have undergone proper IRB review, then potential harms will be minimized, assuming the historian behaves as promised. If historians are not given access to confidential or private information in the first place, then they cannot disclose it, with or without identifying who it belongs to. That is a way to protect people from historical research on stored data, such as that found in archives and manuscript collections (see chapter 4). Then, if a historian conducts IRB-approved interviews or is granted access to stored data (archives) and promises to safeguard and de-identify confidential and private information, but proceeds to disclose all and causes harm, he should be subject to the same possible punishments that unscrupulous journalists are.

There are two problems here for historians. First, is an IRB inappropriately acting as an arm of government by constraining who a historian talks to, how that conversation may be conducted, what sorts of information may be recorded and how identities must be protected, all in ways that ultimately shape what the historian may publish? Second, whose understanding of "harm" should prevail? The scope of possible physical, social, and psychological harms that IRB members (and others, including archivists) might imagine happening and seek to prevent is much broader and more nebulous than the requirements for "harms" that journalists—historians—must be held accountable for in law.

Journalism, History, and the First Amendment

It is not uncommon in conversations about oral history and IRBs for someone to raise the specter of the First Amendment.

> Congress shall make no law respecting an establishment of religion, or prohibiting the free exercise thereof; or abridging the freedom of speech, or of the press; or the right of the people peaceably to assemble, and to petition the Government for a redress of grievances.

Journalists interview all sorts of people without oversight by IRBs, because those they talk to are not considered research subjects. Journalists have frequently published information from stolen or "leaked" documents with impunity, although they may get into trouble if they actually do the stealing, or commit other illegal acts in order to obtain the material. Journalists, in fact, name names and, at times, inflict psychological, emotional, financial, and reputational harm on others. Journalists who pander to sensationalism and indulge in celebrity stalking regularly face criticism—even outrage—from those who find such invasions of privacy in poor taste, if not unethical.

If journalists can talk to whomever they wish without prior review to ensure that their interviews will be conducted ethically, and then publish the details, why can't academic researchers?[2] Surely academic researchers are as protected by the First Amendment as journalists. Indeed, academic researchers have freedom of speech and of publication by virtue of being citizens; some go further and claim that academic researchers have the added protection of "academic freedom," the principle under which colleges and universities resist overt interference with their governance by outside organizations in order to foster as much open debate and inquiry as they possibly can.[3] In counterpoise to this language of rights and freedoms, defenders of IRB review of research using interviews (and secondary data analysis) point out that academics have no fundamental right to perform research. Research, in this view, is a privilege bounded by responsibilities; it is not journalism.[4] It is a legitimate function of government, then, to define those responsibilities and to enforce their application to work that has the potential to harm others, whether physically, psychologically, or socially. The subtle nuance here, nevertheless, is that it is only a legitimate function of government when that government pays for the research or the institutional home of the researcher has volunteered to abide by government requirements.

The arguments on both sides are, of course, much more complex than this sketch can convey. In 2007, the *Northwestern University Law Review* published a special issue containing papers from a symposium, "Censorship and Institutional Review Boards: Getting Permission," in which prominent legal scholars presented positions ranging from seeing IRBs as vital and appropriate protectors of social science research subjects (the possible physical harms of biomedical research were not under discussion) to those who portrayed IRBs, and the regulations that direct them, as instruments of censorship and violators of the First Amendment.[5] Until a case in which a plaintiff challenges a restriction on his research by an IRB on the grounds that the government has trampled on his First Amendment rights reaches the U.S. Supreme Court, and the Justices consider the merits of the arguments, the pro-regulatory position will prevail by default.[6]

No one quarrels with the idea that oral historians, as ordinary people, are free to talk with whoever will agree to talk with them, and to ask them all sorts of questions about their lives. They just cannot do that in a project supported by federal research funds, or as employees of an institution that has signed the general agreement with the Office of Human Research Protections (OHRP) that all research is covered by human subjects protections, and publish the results of their conversations as research, without the possibility that they could get into trouble with their IRB offices.[7] They may, indeed, shoulder the risks to their employment and careers if they wish to do so, and no government agent will stop them. They may be sued by the people they speak with if they publish information their interviewees considered private, particularly if that information was obtained under false pretenses, but that is a risk that journalists and other citizens share. It is in this sense, pro-regulatory scholars point out, that an individual historian's First Amendment right to free speech as a citizen is not abrogated by IRB rules.

The technical debates will no doubt rage on, just as the sense will persist that something, somewhere, is unfair in a society where journalists may ask intrusive questions and name names in public media and academic researchers may not. Some take the line that journalists, known for their "maltreatment of sources," should be subject to the same ethical review as anthropologists. Isabel Awad, a communications scholar, is among these. She dismisses claims that journalism is exempt from IRB review because journalists do not produce generalizable knowledge and so are not doing research. She makes the case that anthropologists who work with a small number of informants also do not produce generalizable knowledge, but, because they can cause harm if confidential material is released, are properly subject to ethical oversight. She goes further and dismisses appeals to the First Amendment as justification for journalists' freedom to publish whatever they wish. For her, the "public's right to know" that journalists equate with First Amendment rights is not a valid position to take, since "not all knowledge is publishable; there is no public right to know about everyone's private life and intimate relationships."[8] She then argues that the Common Rule provides appropriate guidance for "publishing news that the community needs for its maintenance and development" in its demand that benefits be weighed against harms to individuals.[9] Journalists should, in her view, be required to adhere to the Common Rule's standards for informed consent and identity protection. Since the Common Rule does not apply to research on "elected or appointed officials or candidates for public office," the purpose of the First Amendment—to monitor those in power by providing information of public concern to the people—would not be compromised. If journalists had to live up to principles embodied in the Common Rule, in short, they would become more ethically responsible and harms would be prevented.[10]

Arguments that journalists cause harm by intruding into private lives and revealing details merely to satisfy public curiosity rather than to educate people on government actions long predate the National Research Act and the launch of IRBs. One of the landmark legal essays on the "right to privacy" stemmed from press excesses in the 1890s.[11] A series of major U.S. Supreme Court cases in the 1960s through 1980s, however, affirmed the press's power to define what is "newsworthy," and hence protected by the First Amendment, in quite broad terms.[12] Journalists, especially academic journalists who study the profession, have turned to urging journalists to develop and to voluntarily abide by ethical standards in order to prevent the very harms that Awad and so many others have decried.[13] From philosophical analysis in scholarly texts to specific media organizations' codes of ethics, journalist-ethicists urge familiar virtues: avoid conflicts of interest (sources of real or perceived biases); avoid stereotyping by race, gender, or other irrelevant characteristic; seek the truth; avoid harms; respect individuals' privacy and decision-making; obey the law.[14] Journalists, like other professionals, then have to balance these values when they conflict with each other.

A journalist's duty to weigh truth-telling against avoiding harm is particularly relevant here. In their contribution to a collection of essays on journalism and ethics, Deni Elliot, a philosopher, and Daniel Ozar, a health-care ethicist, state this duty in a way familiar to IRB advocates: "will the intended action of the practitioner [journalist] cause potential emotional, physical, financial, or reputational harm?" If so, the journalist must exercise care. She must first be quite sure that the information is relevant and that releasing it treats all of those involved equally and consistently; some must not be unduly favored over others. Only then should the journalist consider the "aggregate good of the community" and the "overall good" being promoted as part of the calculus that justifies possible harm to individuals whose lives intersect with newsworthy events.[15] Academics in law, journalism, and political science, in contrast, stress the press's duty to cover human suffering during natural and man-made disasters, to expose corruption, to provide information on crimes and their perpetrators, and to present details important for democratic decision-making and public policies, even when named individuals might be embarrassed, suffer loss of reputation, or claim intrusions into their privacy.[16]

In their efforts to hold people accountable for their actions by making those actions public, journalists make mistakes. So, journalists themselves may be held accountable. Quite apart from any sanctions imposed by employers, individuals who believe that they have been harmed by journalists—or anyone—who has published their private information have recourse under the civil law of torts.[17] When such lawsuits have been appealed, and so reviewed by higher courts that consider how laws are interpreted, the courts have decided just how accountable journalists must be when exercising their First Amendment rights.

Torts: Invasion of Privacy

Torts are harms. Tort law is the branch of civil law that redresses harms that individuals do to one another when those harms are not actual crimes. For an action to be a tort there must be evidence of real harm, whether reputational, financial, physical, or psychological. Just being embarrassed, upset, or angry does not count. The punishment for being found liable for a tort is damages, not jail time, and can include punitive damages as well as compensation for the actual cost of the harm inflicted. In theory, anyone can bring a civil lawsuit against anyone else claiming that a harm has been done, but that does not mean that a court will take it seriously. The threat of being sued, nevertheless, can have a powerful chilling effect on non-lawyers since it raises the specter of legal fees and time-consuming anxiety. So, how realistic might such threats be for historians?[18]

Publishing private information about another person falls under the more general tort of invasion of privacy. This tort has four broad parts: appropriation of another's likeness or identity for one's advantage, "unreasonable intrusion" into a person's physical space, "public disclosure of private facts," and "false light in the public eye."[19] The first area refers to using someone's image or name to promote a product or service for profit without that person's permission. The second covers electronic surveillance as well as physical intrusion into private space. The fourth is closely related to defamation, but refers mainly to publicly ascribing something to a person (an opinion, support of a political position, a lawsuit) that is true, but in a way that gives a false impression about the person. It is the third area that concerns us here, as the "public disclosure of private facts" is what journalists, historians, and other researcher/authors might do when investigating stories and publishing details affecting living individuals without their consent.

Tort laws may now be grounded in statutes, but they evolved through cases in which juries and judges decided that harm had been done. William Prosser, a major legal theorist, summarized the common law (case law) of torts from the 1940s to the 1970s. He articulated the guiding principles for what constitutes "public disclosure of private facts":

1) the disclosure of the private facts must be a public disclosure and not a private one;

2) the facts disclosed to the public must be private facts, and not public ones; and

3) the matter made public must be one which would be highly objectionable to a reasonable person of ordinary sensibilities.

To this core, he added a fourth requirement that "the public must not have a legitimate interest in having the information made available."[20] The first two conditions may seem obvious, but they cover important ground. A person (assuming the person is not a doctor or attorney acting in a professional role) is not likely to be found liable for invasion of privacy when violating a confidence simply when gossiping over coffee, for example. That counts as a private disclosure, even if the gossip spreads by word of mouth. Nor will a person likely be found to have committed an offense when he publishes the make and model of a car his neighbor drives, since the car is taken out on public roads, making its make and model a public fact. If a fact has made it into a public record, even if that record is a paragraph in a fifty-year-old community newsletter, it cannot be later claimed to be a private fact simply by virtue of the passage of time (see more on this later).[21]

The third and fourth criteria are more likely to inspire spirited arguments by opposing attorneys in a lawsuit. The third qualifies the type of information that can cause harm. "The law is not for the protection of the hypersensitive, and all of us must, to some extent, lead lives exposed to the public gaze," notes the most current edition of Prosser's *Handbook of the Law of Torts*. The threshold for bringing a lawsuit, then, is not the point at which a particular person feels that she has been harmed, but when the ubiquitous "reasonable person" has had "highly objectionable" information made public and has been actually damaged by it. Since the 1960s, the courts have nearly always found for the media when citizens have claimed that the press violated their privacy by publishing information about them in news stories or by showing them in news photographs.[22] Only when journalists have used deceptive means to enter spaces where people have an expectation of privacy, or have used hidden recording devices in such spaces, have courts been less tolerant, but those cases have involved intrusion into private areas as much as publication of private facts.[23]

One of the central reasons why the media has prevailed in most cases rests on the fourth criterion: the public's "legitimate interest" in the information. In the 1960s through the 1980s, the tension between the recognition of a tort of "public disclosure of private facts" and First Amendment claims for freedom of the press came to a head, and courts decided that newsworthiness covered not just stories about politics, the economy, and other serious subjects, but also entertainment and the public's curiosity about the lives of other people. "The scope of a matter of legitimate concern to the public is not limited to 'news'"

stated the authoritative *Second Restatement of Torts* in 1977, "in the sense of
reports of current events or activities. It extends also to the use of names, like-
nesses or facts in giving information to the public for purposes of education,
amusement or enlightenment, when the public may reasonably be expected to
have a legitimate interest in what is published."[24]

As Amy Gajda discusses in her law review article "Judging Journalism,"
however, the attitude that anything goes started to run into pushback in the
1990s as news cycles became ever more intense. "Respect for the media has
fallen," she explains, with the explosion of trivia, scandals, and sensational
stories filling twenty-four-hour cable networks and the internet. Some jurists
believe that it is time to put teeth back into the tort for invasion of privacy,
and have referred to journalists' own codes of ethics as standards that may be
applied when deciding excesses in particular cases.[25] The problem with that
approach, as other legal scholars point out, is that codes of ethics are aspirational documents, not legal ones, and they open the door to allowing juries and
judges to substitute their own feelings about a particular journalist's actions
using various codes' phrases (e.g. avoid harm) instead of making decisions
based on a professional understanding of journalists' rights and duties.[26]

The Passage of Time: Journalism as History

The good news is that historians rarely get sued for invasion of privacy—at
least to the extent that such lawsuits reach a court of appeals. We have to
look towards lawsuits against journalists to find legal reasoning for deciding
whether First Amendment protections apply in particular cases where information from the past has been dredged up and put on display. How seriously, to
put the issue bluntly, have courts taken lawsuits over the "public disclosure of
private facts" when those facts are old ones?

In 1967, Marvin Briscoe sued *Reader's Digest* for, among other things, the
"public disclosure of private facts." *Reader's Digest* had summarized an article
about truck hijackings in which the author named the young Marvin Briscoe as
a beginner at this crime, without saying that Briscoe's act had occurred eleven
years before. Briscoe, who had "abandoned his life of shame and became entirely
rehabilitated," did not deny the truthfulness of the facts or the newsworthiness
of the article's topic. He believed, however, that "his name was not" newsworthy. The article could have appeared without it with no loss of significance.
The trial and district court had dismissed the suit with the basic argument that
the publication of truthful facts was protected by the First Amendment. Justice
Raymond E. Peters, of the Supreme Court of California, however, wrote for the

court's decision to overturn the lower courts' dismissals. Justice Peters decided that it was simply up to a jury to decide the facts of the case: was the plaintiff's name newsworthy? Was it horribly offensive to community sensibilities to reveal a criminal act in the past of a rehabilitated, upstanding citizen?[27] The lower court was required to decide.

E. R. Roshto was equally offended in 1977 when his local newspaper randomly reprinted a page from a twenty-five-year-old issue, which it did occasionally for historical interest. On the front page of the November 14, 1952 edition was a story about his prison sentence for cattle theft. According to Roshto, "the 25-year old matter was no longer of public concern." He never disputed the truth of the report, so the questions, again, were newsworthiness and offensiveness, Justice Harry T. Lemon wrote for the Supreme Court of Louisiana. The court weighed the arguments. "The passage of a considerable length of time after the pertinent event," he wrote, "does not of itself convert a public matter into a private one." He continued: "Lapse of time is merely one of the factors to be considered in determining liability for damages for invasion of privacy by publication of an offensive but truthful matter which was once of public concern and is still of public record."[28] It was important to Justice Lemon that no malice was involved in the act. The page was chosen at random, and it marked the history of the community itself. The court also did not find that there was "any abuse in the purpose or manner of publication." The newspaper publisher was "arguably insensitive or careless in reproducing a former front page for publication without checking for information that might be currently offensive to some member of the community. However, more than insensitivity or simple carelessness is required for the imposition of liability for damages when the publication is truthful, accurate and non-malicious."[29] The publisher won the appeal and did not have to pay the damages that a lower court had imposed.

In 2004, thirty-three years after the Supreme Court of California decided that Marvin Briscoe had a point that the passage of time may have made his name less newsworthy, the court firmly reversed that decision in *Gates v. Discovery Communications.* A dozen years after Steve Gates finished serving a three-year prison sentence for his role in a murder for hire, Discovery Communications created a documentary television film about the crime, which aired in 2001, and showed a photograph of him. Gates claimed that he was harmed by this invasion of privacy, since he had established himself as a law-abiding, rehabilitated citizen and this revelation negatively affected his standing in his community. The court did not agree with his attorney's arguments and, in keeping with other cases that had been reviewed in the intervening decades, decided that publishing facts from a public record—no matter what sort of public record or its age—was protected by the First Amendment. This case law, also enshrined in several U.S. Supreme Court decisions, thus firmly supports historical research using public documents, even if those public documents contain information that living people would very much like to keep suppressed because its public dissemination would cause them distress and harm them in the eyes of others. This is so because the First Amendment protects all citizens, not just journalists.[30]

So far so good. But what happens when the source is not a public document? A key case for historians reached the United States Court of Appeals for the Seventh Circuit in 1993.[31] In 1991, Nicholas Lemann published *The Promised Land: The Great Black Migration and How It Changed America* (Alfred A. Knopf), and it reached the New York Times Best Seller list for non-fiction on April 14. Luther Haynes, the husband of one of the key people whom Lemann interviewed, and an interviewee himself, objected to the details about his life in the 1950s and 1960s that Lemann disclosed from conversations with Haynes's wife, Ruby Lee Daniels. Chief Justice Richard Posner wrote the opinion for the court, which affirmed the district court's summary judgment dismissing Haynes's claims for defamation and invasion of privacy.

The defamation counts, where Haynes asserted that Lemann published false information that harmed his reputation, were dismissed because, even if it were true that some of Lemann's statements were inaccurate, those inaccuracies were trivial compared to the truthful information in the book (and to information from the public record that Lemann could have included in his book, but did not) that shows Haynes in a very poor light. At the time of the events that Lemann narrated, Haynes could not contest that he was a heavy drinker, lost jobs, and refused to support his four children with Ruby after he left her. Lemann did not even mention that Haynes had entered into a bigamous marriage, which a search of public records made clear. "If the gist of a defamatory statement is true, if in other words the statement is substantially true, error in detail is not actionable. . . . Substitute the true for the false (if Haynes is believed), and the damage to Haynes's reputation would be no less."[32]

More important to us here are Haynes's claims for invasion of privacy. Like Briscoe, Roshto, and Gates, Haynes was upset because he had changed since the 1970s. Justice Posner quoted from Haynes's deposition:

> I know I haven't been no angel, but since almost thirty years ago I have turned my life completely around. I stopped the drinking and all this bad

habits and stuff like that, which I deny, some of [it] I didn't deny, because I have changed my life. I take me [sic] almost thirty years to change it and I am deeply in my church. I look good in the eyes of my church members and my community. Now, what is going to happen now when this public reads this garbage which I didn't tell Mr. Lemann to write? Then all this is going to go down the drain. And I worked like a son of a gun to build myself up in a good reputation and he has torn it down.[33]

But, Justice Posner explained, the recent trajectory of case law meant that people could not so easily "bury the past" by claiming invasion of privacy when the revived information came from the public record. The case here differed only because "the primary source of the allegedly humiliating personal facts is not a public record. The primary source is Ruby Daniels."[34] Lemann made it clear that he recounted Daniels's story, among others in the book, to represent and to illustrate the effect that moving north had on individuals as part of a larger transformation of American society, including the effect of economic stress and welfare programs. "Luther Haynes did not ask to be a representative figure in the great black migration from the South to the North," Justice Posner observed. But "people who do not desire the limelight . . . [n]evertheless have no legal right to extinguish it if the experiences that have befallen them are newsworthy, even if they prefer that those experiences be kept private." Neither the passage of time, which mattered to Haynes, nor the fact that some of the statements came from Daniels rather than an official public document, could dent the right that Lemann had to judge Haynes's experiences as "newsworthy."

Here we have a wrinkle. At the start of his opinion, Justice Posner had stated explicitly that Lemann was a journalist who had written a "social and political history" of the migration of African Americans from the rural South to Northern cities between the 1940s and 1970s. "It is not a history as a professional historian, a demographer, or a social scientist would write it," Justice Posner asserted. "Lemann is none of these. He is a journalist and has written a journalistic history, in which the focus is on individuals whether powerful or representative. . . . In the latter [group] are a handful of the actual migrants."[35] Justice Posner did not state that the fact that Lemann was a journalist somehow affected Lemann's status in law, just that it explained Lemann's approach to his subject. (Historians, indeed, might take issue with Justice Posner's notion that professional historians do not write books that include individuals' stories.) Nor does any of the material in Justice Posner's opinion even remotely allude to Lemann's journalistic credentials as key for excluding him from standards that an IRB might have applied to

a project based on interviewing poor African Americans. Nevertheless, Haynes
and his second wife, Dorothy (a very minor character in the book),

> question whether the linkage between the author's theme and their pri-
> vate life is really organic. . . . They point out that many social histories do
> not mention individuals at all, let alone by name. That is true. Much of
> social science, including social history, proceeds by abstraction, aggre-
> gation and quantification rather than by case studies. . . . But it would be
> absurd to suggest that cliometric or other aggregative, impersonal meth-
> ods of doing social history are the only proper way to go about it and
> presumptuous to claim that they are even the best way.[36]

Justice Posner continued with a key point: "If he cannot tell the story of Ruby
Daniels without waivers from every person who she thinks did her wrong, he
cannot write this book."

So, the Hayneses countered, "at least Lemann could have changed their
names." Pseudonyms would not have solved the problem, however, because
those who knew the family well would have been able to identify them from the
details in the book. Furthermore, "Lemann would have had to change some, perhaps many, of the details. But then he would no longer have been writing history.
He would have been writing fiction. The nonquantitative study of living persons
would be abolished as a category of scholarship, to be replaced by the sociological novel."[37] Here Justice Posner referred to other cases where the plaintiffs had
argued that their stories may have been newsworthy, but that certain identifying
details were not. A central example is the case of Dr. Beatrice Gilbert, who sued
the Medical Economics Company and the authors of an article published in the
journal *Medical Economics* about impaired physicians and medical malpractice.
The authors included details of her psychiatric history and her marital problems,
along with her name and photograph, to illustrate the "collapse of self-policing
by physicians" as one of the systemic causes for serious medical errors.[38] Justice
Posner quoted Justice McKay of the United States Court of Appeals for the Tenth
Circuit, who decided that publication of the details, including the photograph,
"are substantially relevant to a newsworthy topic because they strengthen the
impact and credibility of the article. They obviate any impression that the problems raised in the article are remote or hypothetical."[39]

For both Justices Posner and McKay, the context of the private information within a larger argument that had social value was paramount for their
reasoning. Neither Lemann nor the authors of the *Medical Economics* article
were simply revealing identities to satisfy prurient interests. Lemann did not

give his readers a "titillating glimpse of tabooed activities."[40] "The material did not contain descriptions of the Haynes's sexual acts, or bathroom habits; nor did it include nude photographs or other illustrations that the average person would find 'deeply shocking.'"[41] "Painful though it is for the Hayneses to see a past they would rather forget brought into the public view, the public needs the information conveyed by the book, including the information about Luther and Dorothy Haynes, in order to evaluate the profound social and political questions that the book raises."[42] Justice Posner thus emphasized that providing details about real people in the course of making historical arguments serves the public good because they allow the public to assess those arguments in light of unvarnished evidence.

As noted earlier, some courts may now be more likely to hear tort cases about "public disclosure of private facts" without immediately dismissing them under the First Amendment. The ones that have been sent back from appeals courts for trial since the mid-1990s have concerned extreme behavior: gossiping in print about a person's multiple sexual partners, quoting comments overheard from a mother whose son was just murdered after she had refused to speak with journalists, and producing a reality TV show centered on catching child predators by soliciting meetings with fake children, during which the subject of one of the shows committed suicide.[43] Any serious restrictions on the First Amendment stemming from tort cases are highly unlikely, however, given the strength of most court decisions since the 1960s. Such excesses, nevertheless, will strengthen calls for journalists—especially the amateur journalists who write blogs, participate in online forums, and post candid videos on YouTube—to voluntarily respect others' privacy and to resist doing harm.

Wrongs against the Dead

If it is difficult to get damages for the "publication of private facts" about the living, it is even harder to get them for publishing information about the dead. First, there simply is no way to sue for the invasion of a dead person's privacy.[44] The only way to sue is to claim that a living person has been deeply harmed by the publication of information about a dead person who was very close to him or her. Civil suits claiming these harms have been regularly dismissed by appeals courts, largely because it is so hard to prove that a claim about a deceased person can really affect a living person's reputation or status.[45] As discussed in chapter 2, harm to the living relatives of a dead person has been upheld to prevent the release of photographs of the dead person's body under a FOIA request. The only civil suits that have been upheld by an appeals court

in the past twenty years have been similar, in that they all involved recent photographs of the dead person, not just statements of facts about him or her, and
depicted the trauma of death in gruesome ways.[46]

People may sue for defamation of the dead, however, where false information is published that directly harms the living person's reputation.[47] Indeed,
in some states—Nevada, Idaho, Georgia, North Dakota, Oklahoma, Texas, Utah,
and Colorado—deliberately trying to "'blacken' or vilify the memory of one who
is dead" is a criminal act.[48] But, although it is a crime in those states, courts
have consistently refused to allow survivors to collect monetary damages for
their deceased loved one's vilification under civil tort law.[49] The origin of such
criminal laws lies in historic fears for the public peace: saying maliciously bad
things about a dead person might rile her or his relatives and spur them to seek
violent revenge, which perhaps explains the geographical distribution of these
aging statutes.[50] The state, in other words, had an interest in preventing social
disruption and retaliatory violence; the crime was not the personal affront to the
deceased person's relatives, but the possible lawless consequences of posthumous libel or slander. Not surprisingly, prosecutions under laws against criminal
libel of the dead have virtually disappeared.[51] Since proving defamation in civil
court requires evidence that the false information was published with at least
some knowledge that it is false, even if not always with actual malice, then we
can pass by further discussion of defamation cases.[52] Responsible historians do
not make up discreditable information about the dead, or knowingly pass on
falsehoods as true statements; historians who do deserve to be sued.

Historians who research and write with meticulous attention to accuracy
and documentation of their sources have little to fear from threats of lawsuits for
the "public disclosure of private facts" about the dead. Even if a historian is given
very personal materials about a dead person's relationship with her attorney, or
his physician, or her priest, the historian cannot be punished for using it. The
dead person's attorney, physician, or priest, if the source of the material, might
be in for a severe reprimand from his or her professional brethren for disclosing
those private facts, but the historian would not be. That is because those professionals had relationships with the deceased person in which confidentiality was
an ethical duty, while the historian has no formal relationship of trust to the
once-living person.[53] If those professionals are also dead, moreover, and the revelations came from papers they left behind in old trunks, then whoever possesses
the old trunks and their contents has the right to dispose of them as they will,
including just giving them to a historian. In some cases, the contents may pass to
an archive, with or without restrictions on when they are opened for public use.

Conclusion

Lawsuits claiming harm from "public disclosure of private facts" will no doubt continue to be filed and tried.[54] Historians may be sued. The defense that the public has a "legitimate interest" in whatever information is disclosed will, in turn, remain a powerful one, since the harm to society from punishing the publication of truth could be much more damaging in the long run than the harm experienced by individuals to their reputations, financial standing, or psychological well-being. In the end, historians are just as protected by the First Amendment as any other citizen, and so are equally free to be as insensitive as any muckraking journalist. Neither journalists nor historians have much to fear from publishing true statements about dead people, as well, although photographs of violent death for those with surviving relatives—as true as such photographs may be—can cross the line into non-protected speech.

What is left over, now that the legal parameters have been explored, are the ethical issues that shape historians' access to sources and their own decisions on what to make public and what to keep sheltered. While how the law works may be complex indeed, there are at least definite processes through which laws are interpreted and decisions are made. Ethical debates about privacy have no such formal procedures for resolution, at least outside of IRBs. Archivists and manuscript curators have had the challenging task of figuring out how various regulations apply to their records. They have also had to make decisions about researchers' access to information based on donors' conditions for depositing their collections and on their own concerns to protect the privacy of named individuals. Understanding how archivists manage privacy, then, is central to understanding historians' ability to carry out historical research.

Chapter 4

Archivists at the Gates

Historians know that most knowledge about past events and experiences has been irretrievably lost. Most people never made material information-containing items, did not keep the ones that they made, or at some point destroyed the ones that they had kept.[1] Some of what survives may then cross paths with archivists who survey it with critical eyes. As Terry Cook bluntly put it, "a major act of determining historical meaning—perhaps *the* major act—occurs not when the historian opens the box [in an archive], but when the archivist fills the box, and, by implication, through the process of archival appraisal, destroys the other 98 or 99 percent of records that do not get into that or any other archival box."[2] The idea that archivists—those actually entrusted to preserve the past and the present-becoming-the-past—deliberately get rid of records is a painful one for historians. There are many reasons why documents (in whatever media) are erased, shredded, incinerated, or otherwise destroyed. One of these reasons is to protect individuals' privacy. Confidential personnel files, psychiatrists' session notes, raw research data: gone. Sometimes some such items are saved, perhaps left among an author's personal papers or in an administrator's routine files, but later found and closed to users for years or decades. Sometimes some slip through, and then historians get to decide how to use them. Historians may create the past through their archival research and writing, but they do so largely based on materials appraised, accessioned, reviewed, categorized, organized, documented, and made available to them by archivists and curators. Or, to put it another way: historians may create the past, but archivists make certain pasts possible.

From at least the seventeenth century, those responsible for the authenticity of records—and so for the authenticity of the political and religious claims based upon them—have thought hard about the meaning of documentation, and the ways that procedures for identifying, collecting, and preserving documents shape that meaning. Archival theory/the theory of archives, has a long pedigree; its study is not for the faint of heart. From Dom Jean Mabillon's *De Re Diplomatic Libri VI* (1681) to Jacques Derrida's *Archive Fever: A Freudian Impression* (French Edition, 1995) and beyond, scholarly analyses of how we deal with written traces from the past have explored both the pragmatic and the philosophical construction of the archive(s), and their close relations, manuscripts collections.[3] Privacy, however, seems to be a particularly modern and post-modern preoccupation, associated with the rise of the bureaucratic state and its dossiers on individual citizens, the increasing number of collections of personal, rather than state, papers, and, most recently, enhanced attention to individual rights and autonomous decision-making.[4]

After a brief introduction of key archival concepts, this chapter probes the multiple ways that archivists think about and manage privacy concerns.[5] Today's policies and practices will directly shape tomorrow's archives and what historians may expect to find. Tomorrow's policies, if applied retroactively, may reconfigure access to yesterday's records. Again, discussions about privacy for the living spill over into de facto practices about privacy protections for the dead. When dealing with collections that contain information about both the living and the dead, for instance, it may be very difficult to separate out the deceased, whose details technically might not be protected, from those of the living, which may be. As time passes, and more of the living die, who is responsible for determining death dates?[6] Is it easier just to close sensitive parts of a collection for fifty, seventy-five, one hundred, or more years from the date of creation? This is a vexing issue for those who care about second- and third-party privacy, as we shall see. Since archivists have thought a great deal about privacy, the range of archivists' positions on it can only help historians to reflect on how they deal with private information that comes to them freely, in open collections; by accident, with the overlooked document tucked into a file or with poorly redacted items; or fortuitously, through materials given or loaned to them by family, friends, contacts, and other historical actors.[7]

Archives and Archivists, Manuscripts and Curators, Paperwork and Records Managers, and the Out-of-Control Internet

Theorists are anxious to remind us that archives are concepts, not definitive things. They are places, and they are points of view. Archives are centers for cultural memory and of cultural forgetting, through what is lost or hidden in the archives, by the obvious gaps on the real or metaphorical shelves, and by the deliberate destruction of compromising materials.[8] The great European and Asian archives are the descendants of the offices of those who ruled, the working records of taxes, laws, contracts, treaties, and proclamations. "Archives" in this context means the records of power, of the state, which became the foundations of public records, the public business of authorities available for citizens to consult. Institutions, too, such as universities, corporations, and religious bodies, have archives, containing the formal records of administrative functions. These, too, are places of power. In contrast, personal papers, the unpublished remains of reports, books, diaries, letters, drafts, accounts, receipts, scrapbooks, photo albums, and greeting cards, are best lumped together as "manuscripts." Manuscripts may be highly significant, but they usually are not powerful in the way that state records are powerful.

Because official records may look like manuscripts, in the sense of written by hand and never published, and because personal manuscripts may be deposited in places called archives, some technical definitions are called for. In this discussion, I follow general archival conventions and use "archives" to mean the records of states and institutions, produced by offices, not by people (even though people created them). "Manuscripts" means items produced by people as they went about their daily lives, which might include unofficial activities within their official and professional lives. These items were generally never published, although many nearly indistinguishable copies may exist, and they were not instruments of administrative authority.[9] A handwritten copy of the Declaration of Independence that I made in grade school, for instance, which is now in a box of my childhood items, has none of the authority of the Declaration of Independence held in the Rotunda of the Charters of Freedom in the National Archives and Records Administration (NARA) in Washington, DC. The former is non-authentic and worthless as an instrument of power; the latter is authentic and extremely powerful. The content is the same. My copy might end up in a manuscript repository, of interest to some future historian of childhood in 1960s white, middle-class America, but that is highly unlikely (I really should clear out that clutter). In contrast, the document preserved at NARA is an archetype, a symbol of a nation, and a centerpiece of an archive. It is seen as enduring.[10]

While "archivists" manage archives and "curators" handle manuscripts in special collections, "archivist" is frequently used for the person who manages both. Institutions that have separate departments of manuscripts and special collections, such as the Library of Congress, have curators. Subsuming "curators" under "archivists" does some violence to the nuances that distinguish them, but many of their concerns and duties overlap to the extent that "archivist" serves as the more generic term.[11] "Archivist" is used in this chapter, except when "curator" is needed to refer to the particular privacy issues that managers of collections of manuscripts face.

For both archivists and manuscript curators, the creation and context of documents means everything for determining what they are, where they belong, what they are connected to, and what rules (if any) govern who gets to see them. Documents do not need to be true, they only need to be authentic (or, more precisely, be trusted to be authentic); even authentic forgeries and fakes have a significant place in the historical record.[12] From the moment that archivists and curators consider a set of documents for acquisition, they begin a process of shaping what the documents can mean out of what is known about their origins, authors, and possessors. Without that context of meaning, a document itself can mean little. Sometimes that context may be determined from a combination of material form and internal clues: the age of the paper, the authenticity of a signature or seal, a written date consistent with other indicators of age, formulaic phrasing, handwriting style, or digital encoding. Much more often, context descends with the document from its known origins and custody, as imperfect as that knowledge may be. For instance, a sheaf of letters from the late 1880s contained in a file labeled with the name of the sender, which arrived in a box at NARA after having been packed in an office in the Department of Agriculture's defunct Section of Vegetable Pathology, from a file cabinet labeled "Western Botanists," is so defined. A single undated letter from the file addressed "Dear Sir" and signed with an illegible scrawl and no place of origin given, taken out of the file and out of the NARA building, is cast adrift, its meaning eroded.

Over the last fifty years or so, a third specialist has emerged: the records manager. Government agencies, corporations, and other large institutions deal with so much in-house information, both in print and in electronic form, that trained staff people are responsible for keeping track of the paperwork, having taken over this task from the head clerks and secretaries of earlier decades. It is the records manager's job, often in conjunction with information technology professionals, to create and maintain physical and virtual filing systems for the items that are essential to the institution's function, including protecting the documents that must be kept for legal as well as administrative reasons.[13] It is from this vast quantity of daily information flow that, in theory, archivists identify documents that contain information of "enduring value" and pluck them to organize, preserve, and make available to users.[14] Records managers in each agency of the federal government, for example, are supposed to work with archivists at NARA to decide what gets saved at NARA, what is saved in the agency, what becomes classified, and what gets shredded.[15] Records managers, unlike archivists, have no duty to value public accessibility. Indeed, those who manage medical records in hospitals, for instance, have a duty to ensure that the vast majority of people do not get their eyes on the information they control. Historians deal with records managers (or their equivalent), not archivists, when negotiating access to materials still held in active departments, such as when making a FOIA request to the Department of Justice, and when seeking recent information from private companies or institutions. The distinctions between "archivist" and "records manager" are obscured, however, when files that contain restricted materials are transferred from an agency or department to a government or institutional archive for long-term preservation. State archives that hold historically important records from defunct state mental institutions, for example, cannot treat those materials as publicly open records (unless the state allows them to be opened after a long period of time), but may be allowed to manage access to them for researchers.[16]

Over the last thirty years or so, yet another repository of information has grown to intrigue both historians of the recent past and those imagining the resources that historians in the future will have to understand the late twentieth and early twenty-first centuries.[17] This repository (the internet) has no archivist or curator in charge of preserving its contents (World Wide Web pages), although the non-profit Internet Archive (https://archive.org) stores glimpses of some of its past incarnations and various institutions and groups capture, annotate, and keep specific time-stamped pages.[18] While the idea that "everything" is on the internet is patently untrue, as is the belief that once something is on the Net it is there "forever," both myths feed the explosion of worries over personal privacy. Ironically, accessible web content might be increasingly fleeting, heightening historians' fears that future studies of the Internet Age will face gaping holes in this transformative media. For example, some web engineers "studied the persistence of content between 2000 and 2007 and discovered a rate of only 55 percent alive after one day, 41 percent after one week, 23 percent after one hundred days, and 15 percent after a year."[19] Nevertheless, the knowledge that social media companies like Facebook keep copies of everything posted, even when items are taken down from public view, along with the awareness of how much personal information is stored on protected corporate and government servers, means that the potential for unwanted access to individuals' data seem unlimited. In addition to data protection and data disclosure laws (of which there are many), privacy advocates—especially in Europe—have been arguing that individuals should have a "right to be forgotten" in order to protect their current and future privacy. Legislation supporting this claim would allow people to demand that social media sites remove any videos, images, or text about themselves from others' pages, and to require search engines like Google and Yahoo to delete their personal data from "the internet."[20]

Historians who wish to approach the internet as a historical archive face the same ethical concerns as other social scientists who want to mine it for research purposes. Some ethicists have argued that the information that individuals post on many publicly accessible sites should be treated with respect for the posters' privacy.[21] This is particularly the case for blogs whose authors use pseudonyms but who could be identified with a little work, and for conversations on discussion boards where researchers lurk or participate without identifying themselves as researchers. Even when people use screen names, the argument goes, they have not consented to have their self-revelations and personal stories used in research. On March 13, 2013, the U.S. Department of Health and Human Services Secretary's Advisory Committee on Human Subjects Research Protections issued a non-binding advisory letter for IRBs to use when developing policies on reviewing internet research that suggests that IRBs will be increasingly apt to want to review such research, even if only to decide that projects are exempt (see chapter 2).[22] As more of the internet contains the past, historians will have to deal with its decentralized, leaderless, and dispersed archival terrain and the ethical concerns access to it entails. Since most historians still deal with physical archives run by archivists who make decisions about privacy and research access, however, including decisions to create online digital copies of their collections, the rest of this chapter focuses on them.[23]

Professional Values and Ethical Standards:
The Society of American Archivists

Codes of professional ethics are useful aspirational documents. Couched in general terms, they provide principles that help to present the profession to the public. They can be used for administration and for advocacy. The Society of American Archivists (SAA) is the largest professional association for archivists and curators in the United States, and it works closely with similar organizations in other nations. The SAA has a formal list of "Core Values" and a "Code of Ethics."[24] Ideally, these declarations should help archivists with decision-making. Like many other statements of professional ethics, however, they embody values that must be balanced when figuring out how to act in specific circumstances, and they provide little guidance for how that balance is to be reached.[25] In particular, archivists must balance "Access and Use" with "Privacy" as they go about their daily business. The passage on "Access and Use" in the Code of Ethics proclaims:

> Recognizing that use is the fundamental reason for keeping archives, archivists actively promote open and equitable access to the records in their care within the context of their institutions' missions and their intended user groups. They minimize restrictions and maximize ease of access. . . . [Archivists] work with donors and originating agencies to ensure that any restrictions are appropriate, well-documented, and equitably enforced. When repositories require restrictions to protect confidential and proprietary information, such restrictions should be implemented in an impartial manner.[26]

Access is clearly valued in the context of responsibility to restrict information to those authorized to use it. At the same time, archivists value the overarching principle of justice: no users should be arbitrarily given access when a similar user is denied. So far, so good.

Privacy concerns are similarly articulated in the Code of Ethics:

> Archivists recognize that privacy is sanctioned by law. They establish procedures and policies to protect the interests of the donors, individuals, groups, and institutions whose public and private lives and activities are recorded in their holdings. As appropriate, archivists place access restrictions on collections to ensure that privacy and confidentiality are maintained, particularly for individuals and groups who have no voice or role in collections' creation, retention, or public use. Archivists promote the respectful use of culturally sensitive materials in their care by encouraging researchers to consult with communities of origin, recognizing that privacy has both legal and cultural dimensions.[27]

Archivists, in short, have both legal and ethical duties to respect privacy. To carry out these duties, archivists have the responsibility to know what is in their collections and the authority to control access to them. They may impose more restrictions than required by law, applying their own assessment of what protections are necessary and appropriate.

The SAA leadership urges that these ethical principles be interpreted in light of the organization's "Core Values." One of these core values is "social responsibility."

> Underlying all the professional activities of archivists is their respon-
> sibility to a variety of groups in society and to the public good. Most
> immediately, archivists serve the needs and interests of their employers
> and institutions. Yet the archival record is part of the cultural heritage
> of all members of society. Archivists with a clearly defined societal mis-
> sion strive to meet these broader social responsibilities in their policies
> and procedures for selection, preservation, access, and use of the archi-
> val record. Archivists with a narrower mandate still contribute to indi-
> vidual and community memory for their specific constituencies, and in
> so doing improve the overall knowledge and appreciation of the past
> within society.

The duty that archivists have to "their employers and institutions" is a perva-
sive theme throughout the SAA Core Values and Code of Ethics. Sporadically
since World War II, and with increasing determination since the mid-1990s,
archivists have also emphasized that they have a responsibility for the "public
good" that, at times, might transcend loyalty to their institutions and employers. Some of the most prominent examples of such activist archivists include
individuals like Verne Harris, a South African archivist who worked to prevent
the destruction of materials by the apartheid government during the transition
to democracy in the early 1990s.[28] Harris has been a loud voice for the social
justice movement among archivists, a movement that explicitly recognizes the
archive as an instrument of power. Social justice archivists seek to unbalance
that power by, among other actions, advocating for open access to government
files, and gathering materials from underrepresented minorities and other marginalized social groups to counter the elitist bias of all established archives and
most special collections.[29]

A major U.S. example of archivists acting quickly for the public good
occurred in 1994, when Stanton Glantz, a professor in the University of California at San Francisco (UCSF) medical school, received a large box of copies
of documents stolen from the Brown and Williamson Tobacco Corporation. He
offered these documents to the UCSF archive and Karen Butter, the director

of the USCF archive and library, accepted them. She had the support of UCSF legal counsel, a necessity when Brown and Williamson subsequently sued. Butter did not act against her institution's wishes in this instance, but she certainly took on a large burden of time and emotional energy by attracting the ire of large corporations that had a strong claim that the stolen documents contained protected proprietary information.[30] For Glantz, Butter, USCF attorneys, and, ultimately, the courts, the public good promised by the revelation of corporate officials' deceptions about the health risks trumped the legal and ethical rights that the corporation had to regain control over its property.

Examples of American archivists as agents of resistance and exposers of crime, corruption, fraud, and ethical breaches are hard to find and even harder to document.[31] More obvious are those archivists, like Karen Butter, who use the opportunities provided by their institutions' collecting goals to acquire, process, and provide access to potentially controversial materials. The SAA's Core Values and Code of Ethics are formal reminders of the expectations for professional, ethical behavior among archivists: providing access, protecting privacy, and, where possible, seeking the public good. Nevertheless, in the end, historians must appreciate that archivists have a primary duty to serve the needs of their employers and institutions. That duty might include the diligent preservation of evidence of wrongdoing, in hopes that someday that documentation might be used to aid in the redress of harms done.[32] That duty might also include gut wrenching decisions to refuse to accept collections that are simply too politically hot to handle.[33]

Archivists are called upon to assess legal and ethical privacy concerns at several points when they deal with materials. First, archivists have to decide what to acquire. That may involve articulation of a formal set of collection goals, or it may be done on the fly. Second, once acquired, materials must be processed in order to get them ready for use. Third, once processed, access conditions must be set. This nice set of steps is somewhat of an illusion, of course, since decisions about acquiring a collection are linked to whether the collection may even be used in the next thirty, fifty, or one hundred years. As boxes of papers or gigabytes of files are gone through and itemized, sections might need to be set aside because their content makes use problematic. Dividing an analysis of privacy concerns into the stages of acquisition, processing, and access, however, at least offers an orderly approach to what archivists do well before history can possibly get written.

Appraisal and Acquisition: Privacy Exclusions

What gets saved? This is a key question facing records managers, archivists, and manuscript curators on a daily basis. We are overwhelmed with information and have been increasingly inundated since at least the 1960s. We know that only an exceedingly small fraction of the inscriptions we now make can—or should—be preserved for decades. Those well-versed in archival theory know perfectly well that what they do literally constructs our multiple collective pasts when they decide what gets saved and what gets destroyed or recycled.[34] They know that the archive is a foundation for power. They know the myriad ways that past voices have been silenced by exclusion and destruction. Those with historical training, too, can all too easily imagine how a future historian might use just about any set of materials, from fast food menus to the blogs of incipient revolutionaries. The democratic impulse, the historical impulse, is to save everything so that no one is inadvertently denied inclusion and that no future histories can never be written.[35] This is impossible, both pragmatically and philosophically. So, records managers, archivists, and manuscript curators must choose from the wealth of materials which come to them from their institutions' offices and from individual donors.[36] This is not easy to do.[37]

Let me begin with a fairly straightforward example. One of the duties of the archivists at NARA is to appraise the records of federal agencies in order to determine which have historical merit and to put them on a retention schedule. At the same time, NARA has established various policies to cover the routine destruction of materials that do not need to be saved. Such forward planning makes managing the current avalanche of information much easier. The need to protect individuals' privacy is a clearly recognizable selection tool. So, one of NARA's hundreds of regulations (plucked at random) states that the staff of most federal agencies are to destroy all "correspondence, reports and other records relating to interviews with employees" six months "after transfer or separation of employee [sic]" unless those records were created before January 1, 1921, when NARA was organized.[38] From this perspective, documents have a life cycle, from origination to destruction, and the more that archivists work with records managers to decide at origination which materials are to have a long life span, the better the government's, institution's, or corporation's records can be kept in proper, tidy, and manageable order. Some items get saved; all of the rest may be tossed. While the archivists are supposed to be the final arbiters of appraisal and disposal, however, they have little power to police what goes on inside offices, and records certainly have been destroyed without informed consultation.[39]

Of course government agencies, institutions, corporations, and other large entities have quite a bit of personally identifiable information about citizens, employees, clients, customers, students, and patients accumulate in their files, much of which is already covered by privacy laws that in themselves restrict what is likely to be saved for any length of time, much less transferred to a repository to which the public has access. Since much of this information is increasingly kept in electronic files, records managers have suggested that one way to save information for research use is to anonymize it, as the federal government does with aggregate data in various agencies. This idea, appealing to many, mirrors common research practices in the biomedical and social sciences. As discussed in chapter 2, there is ethical concern over using any information gathered for one purpose for a secondary purpose without the consent of the subject. Hence, raw data collected for administrative or research purposes are regularly destroyed, stripped of key identifiers for limited secondary research access, or anonymized for public access.[40] It seems straightforward then, that such de-identified (a selected set of identifiers removed) or anonymized (all fields removed that can allow unique clusters of data to possibly lead to re-identification of individuals) information could be transferred to archives where, after a suitable period of time to reduce even statistically improbable re-identification, it could be made publically available without any concerns for violating individual privacy. The problem, as some thoughtful archivists have noted, is that once identifiers are permanently removed, that is it. The information becomes decontextualized and loses authenticity. The unique individual records ultimately become inaccessible for any later use to redress injustices and to provide counter-narratives to those produced by dominant political, economic, and social powers. As Malcolm Todd put it, "the collective droit de mèmoire [right to remember] is effectively undermined by the exercise of the individual droit de l'oubli [right to forget]." From this perspective, long-term archival and historical interests require that only copies of data be de-identified or anonymized, and that the original files be preserved without manipulation for future use. If such data is to be saved at all, advocates assert, save it properly.[41]

Thinking about the archival preservation of huge data files only adds to the burdens that archivists face when trying to decide what should be saved for posterity. Which items, which files, really have historical merit? According to some archival theorists since the mid-1980s, that is not an appropriate question at all. Instead, they argue, archivists need to decide what must be documented about events, processes, and issues relevant to their own institution's context, be that the workings of a federal or state agency, a corporate division, or a university administration, and leave historical relevance to future historians.[42] This plan works reasonably well (in theory) for in-house records managers and archivists, but it is far harder to apply to institutions whose primary purpose is to collect materials for historical preservation and use.[43] This is largely the realm of public and private archives and libraries who accept, by donation or purchase, manuscripts relevant to their institutions' missions. Some collecting goals have been written in exceedingly broad terms, although they may not have seemed that way decades ago. The Bentley Historical Library at the University of Michigan, for example, preserves both university materials and the "Michigan Historical Collections," established in 1935 to "document the history of Michigan and the activities of its citizens and institutions."[44]

Such "collect everything" policies, as Timothy Ericson has put it, were "written more to legitimize collecting activity than to focus it [and] have not been able to save us from ourselves."[45] Archivists need to be "saved" from the random, undiscriminating collecting born from the belief that nearly everything has potential historical value, and the desire to have those things for their collections. Instead, such theorists argue, archivists and curators should develop well-defined parameters for what they will consider acquiring, be selective about materials that duplicate existing holdings, and advise donors about other repositories more suited to the items they have on offer. The collecting policy developed in 2008 by the American Heritage Center at the University of Wyoming has been held up as an exemplar of wise planning after years of indiscriminate collecting.[46] Here, too, privacy concerns are evident. The American Heritage Center, for example, "generally does not acquire" either "personal financial records including checks or income tax returns" or "medical records."[47] How often archivists actually turn down collections because it is too hard to manage privacy restrictions, or because the private information in them is simply too sensitive, is unknown. According to at least one senior archivist, such rejections are very rare—archivists do want to preserve historically valuable materials—but they do happen.[48]

As wise as it may be to refuse items that seriously compromise individuals' privacy, it is hard for archivists to do so in advance. Dealing with sensitive items, especially items that might seem to invade the privacy of unwitting third parties, is then deferred to the next stage of archival work: getting the newly arrived boxes of stuff processed so that they can be opened to users.

Processing: Protecting Privacy by Closure and Redaction

Before a collection of archival or manuscript materials can be accessed by users, archivists must get it ready. At their most painstaking and thorough, archivists take everything out of the boxes the items came in; arrange the items in proper chronological and topical order; remove all paperclips, staples, and other fasteners that might corrode; examine each item to see if it needs further preservation or needs to be put aside for special consideration due to sensitive content; refolder the documents (if they even came in folders to begin with) into archival quality, acid-free folders with hand-written labels (no adhesives); make a list of all of the folders; put the folders into archival quality boxes; label each box with its contents; and create a comprehensive description of the collection. Curators separate out materials that a donor has asked be closed for a period of time, sometimes even for years after his death; archivists sequester information about identified individuals in government documents (except for those who created the documents, who are public figures).[49] Finally, information about the collection must be correctly entered into online catalogs so that searchers may discover it. This entire process takes a long time and it is often expensive.

According to data collected in 2003–2004, 34 percent of American repositories had "more than half of their collections unprocessed; 60 percent [had] at least a third of their collections unprocessed."[50] Some archives are years and years behind with getting materials ready, even as they seek further donations. In 2007, Mark A. Greene and Dennis Meissner made a strong argument that archivists had to let go of their obsession with detailed processing and make collections available much more rapidly. While there will always be exceptions for highly fragile materials, or those that come with legal or donor restrictions, most collections only require minimal review and cataloging. In particular, they noted, "we must get beyond our absurd overcautiousness that unprocessed collections might harbor embarrassing material not accounted for in deeds of gift."[51]

The fear that sensitive items might be inappropriately released is one justification that some archivists give for the need to examine each item before declaring it fit for public use. A processing guide from 2001, for example, suggested watching out for "sensitive subjects as adultery, alcoholism, drug abuse, homosexuality, lesbianism, mental illness, or suicide."[52] An archivist at the Leo Baeck Institute at the Center for Jewish History in New York noted that "sensitive" information for them included hints that individuals were

collaborators with the Germans during World War II. This was of particular
concern because relatives donate papers written in German, which they do not
know how to read.[53] Donors, or their families, should be alerted to such possibly stigmatizing information, many curators believe, and be given the opportunity to have the items returned to them.[54] Even without consulting donors,
especially if the donor was an institution and not a family—as can happen with
faculty papers—archivists have removed uncomfortable items. In one case, for
example, an archivist permanently removed the original suicide note of a well-
known academic that came in with the professor's papers, deciding that it had
no relevance to his life's work.[55]

When sensitive (however defined) information involves second or third
parties, not just details about the donor's life, archivists are divided on what
should be done, especially if the donor did not care about others' privacy or
reputations. "Second" parties usually refer to authors of letters to a donor,
or similar items; "third" parties are those whose names and information are
mentioned in the donor's diaries, letters, or other personal papers.[56] Marybeth
Gaudette calls both of these parties "blind-donors" to emphasize that they have
unknowingly had personal information about themselves potentially made
available to anyone without their consent.[57] Since people disagree over what
constitutes sensitive information, some archivists argue that users (journalists,
genealogists, historians) should be left to decide what is ethical to reveal.[58] Still
others propose that laws be passed to require archivists and curators to contact
all blind-donors to alert them of the materials donated to them. Blind-donors
could then consent to the release date the donor mandated in his or her deed
of gift, or demand longer closure. Yet still others urge that there be mandatory
periods of closure for all information about identifiable individuals. Gaudette,
for example, suggests that access dates be in line with copyright protections
on publishing others' work: life of the author plus seventy years, unless both
the donor and blind-donor—if still living—consent to earlier release.[59] Heather
MacNeil, author of *Without Consent: The Ethics of Disclosing Personal Information in Public Archives* (1992/2001), acknowledges that such mandatory
post-mortem closures in effect extend privacy rights over information to the
dead, which, in her opinion, is as it should be.[60]

Those who argue for solutions embodied in law seem to find these appealing because the alternative is the status quo, where it is up to the individual
archivist or curator to decide what to do in light of donor agreements, her own
understanding of professional codes of ethics (not particularly helpful), and
her personal ethical leanings. For Gaudette, "the possibility that two different

archivists or curators could reach different decisions on the accessibility of a given item is in itself sufficient reason cases-by-case decisions should not be tolerated."[61] Much better, in her view, would be mandatory rules, uniformly applied. Whether Gaudette will have her wishes fulfilled is impossible to predict, but that prospect is worrisome, if only because such rules would seriously hamper work on the history of the recent past by closing down much research on the still living (for a critique of applying regulatory oversight to archival materials, see chapter 6).[62] In the meantime, archivists and curators continue to make independent decisions about how they deal with blind-donors, privacy, and concerns over sensitive items, and their decisions clearly affect what items become available to researchers.[63]

When a collection is processed and sensitive items are removed from their original places in their folders, or folders from their boxes, or boxes from the open shelves, the archivist has a choice to make. Should a place holder be left, indicating that something has been removed, and will be restricted for a period of time? This has the advantage of alerting a researcher to gaps in the records that he is consulting, so that he knows to acknowledge them in his work. It also has the disadvantage of alerting a researcher to gaps in the records he is consulting, so that he becomes irritated at the archive and, perhaps, quite determined to discover what has been hidden.[64] Sometimes information about gaps comes in mysterious ways. Judith Schwartz has recounted how, in the late 1970s, Jonathan Katz, author of *Gay American History*, was tipped off about an unmentioned box in a collection at the Minnesota Historical Society by Barbara Gittings, who had received an anonymous note. The Minnesota Historical Society had listed a collection of Whipple-Scandrett family papers as containing nine boxes. An unlisted tenth box contained letters between Rose Elizabeth Cleveland, President Cleveland's sister, and Evangeline Simpson, who later married Henry Whipple (all deceased), which revealed their lesbian relationship. When Katz queried the Minnesota Historical Society, he was told that the box "had been closed until 1980," although who decided that the ostensibly sensitive items should be sequestered is not clear.[65] By leaving the existence of the box out of the finding aid, the archivists could ensure that researchers would not even have even a tantalizing hint of this interesting correspondence. Careful omissions in collections' descriptions or the use of delicate euphemisms ("female companions" instead of "prostitutes"), especially in the searchable texts of online finding aids, are still among archivists' subtle tools for managing the existence of troubling items.[66]

Even archivists and curators who are strongly in favor of opening as much material to researchers as soon as possible face difficult decisions when processing actual collections instead of considering policies in the abstract. In 1997, for example, Julie Herrada sought the papers of Ted Kaczynski, commonly known as the Unabomber, for the Labadie Collection at the University of Michigan, a collection that focuses on the documentation of radical politics in America. Kaczynski at first wanted everything closed until 2020 or his death, whichever came later, except for anonymous letters written to him, or those from the media to him, which could be opened immediately. Herrada rejected this request as too long for the level of interest in his materials, and they worked out "a series of options from which [they] could choose" as the items were processed. Once she began reading the unsolicited letters written to him from people all over the world, however, Herrada discovered that her firm adherence to the value of open access could not stand in the face of the personal details revealed by those seeking some contact with the person whom they imagined Kaczynski to be. "My gut reaction was to close this collection for a long time. . . . I was genuinely worried about the letter writers" she wrote in an article on her decision-making process. "What they did not know was that I was reading their letters and intending to make sure that many others would read them as well. Suddenly, I felt worse than a voyeur. . . . I felt the weight of the world was on my shoulders. I felt like giving all the letters back. I certainly did not feel entitled to them."[67] In the end, Herrada decided that the only way to balance the need to allow researchers access to the correspondence and to protect the letter writers' privacy was to create redacted copies. Every letter was photocopied and names, addresses, telephone numbers, and other place names were removed; the originals are closed until 2049.[68]

In this case, privacy concerns forced the library's staff not only to review and refolder each item in part of Kaczynski's collection, but also to spend the time and resources necessary to protect blind-donor identities by redaction. MacNeil and Gaudette, both staunch advocates for privacy for archives and manuscripts, could reasonably ask the University of Michigan why those who wrote to Kaczynski deserve such care, while those who wrote to Chellis Glendinning, "a pioneer in the field of ecopsychology," do not.[69] Open access advocates would, in turn, likely note that Herrada was actually following the donor's initial wish to have the items closed, albeit only to 2020, not 2049, rather than actively imposing her beliefs on the collection. Donors, in Mark Greene's view, have the best sense of which parts of their collections should be closed to protect second- and third-party privacy, and have a much stronger right to make decisions about items than archivists do. "If we err," he urged, "let us err on the side of access."[70] The Kacznyski case also illustrates the pragmatic significance of institutional resources. Herrada's department could pay for a suitable compromise between access and privacy for materials that she believed should be available to users because they are an important source for understanding the meaning of Kacznyski's place in late twentieth-century radical politics. Other repositories might not be so flush; other collections might not seem so important.

Even after archivists decide during processing that possibly sensitive items should be opened, users may disagree. Waiting for someone to complain that materials should not be available may be a rather passive approach to protecting privacy, but it has the advantage of responding to real, rather than hypothetical, fears.[71] Those most likely to demand that items be restricted, if not destroyed, are—not surprisingly—donors' relatives, the individuals named in donors' diaries, letters or other personal papers, and the relatives of those individuals. When the donor and the named individuals are dead, there are no legal grounds for acceding to such requests (unless there are ownership or copyright interests at stake), but that might not deter family members from launching a lawsuit. Elena Danielson, retired from her position as archivist at the Hoover Institution at Stanford University, noted that it is very hard to estimate the direct impact that such lawsuits have on archives, as they are almost always dropped or settled before reaching court. From her experience, nevertheless, she believes that "the threat of lawsuits damages the historical record more than the actual lawsuits," because they make archivists—and administrators— risk averse. And the more risk averse the archivist, the longer he will restrict potentially problematic collections.[72]

Authors of ethical advice for archivists thus generally recommend that curators take privacy concerns very seriously, if only to avoid negative publicity and threatened litigation. Unless the items have already been widely used by other researchers, they suggest that the contested materials be restricted for a period of time.[73] When the donors and named individuals are living, then technically the curator should refer the complainants to the donor, who is legally responsible for making the items public and who did not insist on a period of closure in her deed of gift. The cautious may simply prefer to remove items of concern to the person who complained and negotiate with him about appropriate restrictions.[74]

Instead of protecting privacy by closing collections for years, or by redacting identifying information from items, archivists and curators may decide to leave the ethics of disclosure to the user. If only users could be trusted to make the right decisions . . . .

Access: Managing Privacy by Managing the User

Get enough historians together in a discussion about their experiences using archives and manuscript collections, and one or more will inevitably tell a story about being denied access to materials because she did not have the right connections, did not convince the curator of her academic credentials, or simply for no reason at all that she could fathom. Such stories are not as common for American collections as they are for repositories in other parts of the world, but they do circulate. In the hallowed past of European archives, access was by permission only, as the records of state were not considered public records. Only in the late eighteenth and nineteenth centuries did public record repositories start to become more widely open to citizens and researchers, and only in the second half of the twentieth century have archivists expressed a commitment to equitable and open access wherever possible.[75] Still, even NARA requires users to present government-approved identification before being allowed into the reading rooms, which excludes those who lack one from consulting the public records of the federal government.

Posting digital copies of archival and manuscript materials on the internet offers the most open and equitable access to them, as well as serving to protect fragile materials from hands-on users. As noted earlier, the explosion of information available on the internet has made many Americans aware of privacy issues in new ways, however, and has heightened demands that personal privacy be protected, although exactly how that is supposed to work for self-disclosures has quite a few theorists scratching their heads.[76] The growth of online access to public records and to photographs of public spaces, moreover, has confronted citizens with the literal meanings of "public" records and public life. Before the late 1990s and early 2000s, for example, property tax assessor files and local court documents were public records that anyone could see on a visit to a county court house, but when they became easily searchable with a simple internet connection people were outraged that anyone could learn the value of their real estate or discover that they were being sued.[77] The first step in managing user access to archival and manuscript materials, then, is not to make any digital copies available on the internet and, as noted earlier, to be very careful in how items are described in online finding aids. This leaves such materials in what Arminda Bepko and others have called "practical obscurity." Privacy is supported not because the information has ceased to be accessible to the public, but because few people have the knowledge or resources to actually find it. As Bepko argues, this profoundly reshapes what it

means for information to be "public," since it limits public access to a particular medium, to paper or in-person electronic interfaces, and not to an internet
server.[78] Only those who can visit the paper or the restricted electronic files
may use them, which supports a de facto informational exclusivity at odds with
the democratic ideal of the "public" record. As anxious as archivists and curators might be to make their non-governmental research collections accessible
to users all over the world, those who are nervous about donor and blind-donor
privacy may well opt for making users show up in their reading rooms.[79]

Upholding the principle of equitable access seems an obvious ethical
stance for those in charge of public records in public archives. It falls apart,
however, for public records restricted by statutory or regulatory law for reasons
of privacy, such as the registers of defunct state mental hospitals and adoption files, even when those records are held in public repositories.[80] It also is
harder for curators of manuscript collections to follow uniformly. In practice,
just because one person has access to materials does not necessarily mean that
everyone should have access. A famous person might decide to allow only
an authorized biographer to study his manuscripts for a period of time, for
instance, which may be acceptable to curators seeking to acquire important
collections.[81] Donors may be the gatekeepers for parts of their collections that
they want restricted as well. They are willing to consider access on a case-by-
case basis and, after a while, as Mark Greene has observed, some donors "have
just gotten tired of us forwarding requests" and have ended up lifting all or part
of the restrictions.[82] Some private repositories, such as the Huntington Library
in California, only admit faculty members, doctoral students, and independent
scholars, who must provide two letters of reference attesting to the quality of
their scholarly work.[83] The idea that only certain people may be trustworthy
users of certain items smacks both of responsible stewardship and of discrimi-
natory elitism.[84]

Replace "restricted files" and "private papers" with "research data," however, and another view emerges. Here is where the "researcher" becomes the
legitimately privileged citizen. The researcher's probity may be judged, as at
the Huntington, by academic status and reference letters. The administrators of
the Minnesota Historical Society Library take a slightly more plebian approach.
There, an applicant for access to restricted records must provide a "detailed
description" of his project and "identify the public interest that outweighs the
pertinent privacy interests." At the same time, the applicant must "respect the privacy and confidentiality rights of all individuals." He must also, "if required . . .
compile only summary data on individuals and not record or disclose the names

of any individuals or any information that would lead to the identification of individuals." The applicant must acknowledge, moreover, that he "may be personally liable for legal action" for disclosing data, committing libel, or violating copyright.[85] While there is a line for the applicant's institutional affiliation, the form focuses more on vetting the research project and alerting the user to his legal responsibilities than it does on determining his academic qualifications. It is then up to the library staff to decide if access may be granted.

Requiring researchers to apply for access and to sign forms on which they promise to protect individuals' privacy and to be liable for the consequences if they do not might seem to be an ideal way for archivists and curators to manage their privacy concerns.[86] It certainly works for state archives like those in Georgia and Minnesota, where research access is built into public records laws.[87] It is more problematic for archives in states where records laws and privacy laws do not give the archives explicit permission to do this. Yet, the move to shift responsibility for privacy protections to users has an obvious parallel in the way that users shoulder the responsibility for respecting copyright protections. Archivists and librarians have generally been quite happy to make sure that copyright notices are prominently posted, stamped on any photocopies, and included in directions for users, as all of these steps mean that the institution is not liable for copyright violations. If a user does violate copyright, then the copyright holder may sue her for unauthorized use and, in some cases, for damages.[88] Similarly, then, if a researcher appears to have invaded someone's privacy by publishing personal information, the affected individual may sue him. That is a legal solution, but it is not, for some archivists, an ethical one, especially if the affected person is not the individual named, but a relative. If the individual is dead, relatives have even less chance of redress.[89]

There is yet another way for a researcher to gain privileged access to restricted information: by establishing trustworthiness via an Institutional Review Board (IRB), or other authorized review committee. IRBs have long been in the business of determining who may access information for research purposes (see chapter 2). These committees are all about determining a person's qualifications for performing research, the appropriateness of her research methods, and of course, her plans to protect living individuals' identities when gathering and analyzing data or when using data collected for other purposes. While archivists and curators are increasingly discussing the possibility of turning to IRBs as avenues for access to restricted materials, there is little evidence that they have implemented any formal requirements for doing so, with one significant exception: research using hospital and medical records.[90]

When the federal Department of Health and Human Services implemented the Privacy Rule for the Health Insurance Portability and Accountability Act (HIPAA) in 2003, those in charge of several historical medical collections found themselves subject to its regulations. Until its revision in 2013, the HIPAA Privacy Rule applied to all medical information produced or stored in covered institutions (i.e. those subject to the Rule) retroactively and in perpetuity; in 2013, the Rule changed to cover information only for fifty years after the death of the patient or client.[91] The HIPAA Privacy Rule was certainly not the first set of regulations that put privacy protections on hospital and medical records, as a patchwork of state laws had already done so, especially for the records of state mental institutions and state-run health care services. It was, however, the first to explicitly protect all health care information about the dead. The HIPAA Privacy Rule also built in ways for people to access health care information without patients' consent for research purposes by going through IRB review. Since IRBs only deal with living people, the Privacy Rule also allowed covered institutions to establish a new review committee, a Privacy Board. These boards, which act much like IRBs, can approve research projects that only need existing medical information about living and dead people.[92] The Privacy Rule includes several requirements that limit the amount of information released for research purposes. If, in fact, a researcher (or the institution) can de-identify the data by removing eighteen specific identifiers (name, addresses below the level of a state or specified large population, all dates more specific than a year, and so on), then the result is anonymized data that may be made public. If the researcher wants names, specific dates, or local addresses, however, he has to describe why he needs them and how he will protect the data from the risks of further disclosure.[93]

Major covered research collections in the history of medicine, notably the Alan Mason Chesney Medical Archives at The Johns Hopkins Medical Institutions and the Archives and Special Collections in the Augustus C. Long Health Sciences Library at Columbia University, offer important examples of ways that the HIPAA Privacy Rule has been applied to historical materials. The Chesney Archives relies on a Hopkins Privacy Board to vet research applications; the 2014 application for historical research there is eight pages long, with an additional four pages of legal information attached.[94] Those who wish to research collections that contain protected health information about either living or deceased individuals held in the archives at Columbia must complete the "Investigator's Certification for Research with De-identified Data" and sign a confidentiality agreement. The archivist reviews the forms and makes a decision about access. Since the HIPAA Privacy Rule went into effect in 2003, Stephen Novak, the archivist at the Augustus C. Long Health Sciences Library, has only had to turn down two requests, both for projects involving the publication of names in practitioners' casebooks from the 1810s–1840s. These are requests that would likely be granted now, as all of the individuals named in them have clearly been dead for over fifty years.[95]

The overall effects of the HIPAA Privacy Rule on historical research are impossible to assess, since we have no way to estimate how many historians give up on a project before even starting it out of concern that it will be too difficult to access old health care documents. One intangible effect, nevertheless, seems to be a belief that the original Privacy Rule, which closed "protected health information" in perpetuity is somehow more ethical than the revised one, which only applies to health care information for fifty years after patients' deaths. Do note that the Rule does not say that health records are to be opened at that point, only that administrators need not worry so much about strict compliance. Another effect, moreover, has been the voluntary adoption of Privacy Rule-like restrictions on collections that are not technically covered by HIPAA—which are the vast majority, since so few are parts of institutions that provide or pay directly for health care—or by special state laws for medical privacy.[96] In their important 2015 report, "Recommended Practices for Enabling Access to Manuscript and Archival Collections Containing Health Information about Individuals," for example, Phoebe Evans Letocha of the Chesney Archives at Johns Hopkins and Emily Gustanis of the Francis A. Countway Library of Medicine at Harvard begin by urging repositories "to recognize individually identifiable health information" regardless of whether or not they are "entities covered" by HIPAA.[97] If the federal government determined that health care information in some institutional contexts should be protected, even for long-deceased patients, then voluntarily protecting health care information in all repositories can seem to be the most ethical position to take.

If archivists and curators (or the higher-up administrators) of non-covered institutions choose to adopt Privacy Rule-like protections on items that obviously contain health information, such as a doctor's case book from the 1920s, then it remains an open question on whether they will also choose to implement ways for researchers to access that information while promising to protect patients' identities—at least before every named person has been dead for fifty years. They could take the time and effort needed to establish procedures to review research proposals. They could ask researchers to sign confidentiality agreements. They could inspect all notes taken of restricted materials. They could make copies of documents and redact all eighteen of the HIPAA identifiers (although it is hard to do that for location, presumably already named in the collection's description) to anonymize the data. All of these steps enshrine a process for protecting the medical privacy of blind-donors that could be adapted for the use of other identifiable details in all sorts of records. All come with additional burdens on staff responsibilities and staff times. None of them can ensure researcher compliance. Yet, neither can all the federal regulations and IRB reviews in the world ensure researcher compliance. Research in the biomedical and behavioral sciences proceeds because we are willing to accept the risks to gain the potential benefits. Let us do the same for history, at least by allowing current ad-hoc, minimal-risk practices to continue until there is evidence that historians cause real harms to living people greater than those experienced during the ordinary embarrassments of daily life.

Conclusion: Archivists at the Gates

Archivists have a great deal of power over what can be known about the past. They face hard decisions when appraising, acquiring, and processing new collections. Concerns about individuals' privacy are always present, albeit much more intensely at some times than at others. They regularly make difficult judgment calls. Ideally, all items will be available to users—but not always now or in ten or twenty or fifty or one hundred years. In some archives and manuscript collections, items are either open or closed. In others, staff take the time to redact names and other identifiers from copies of documents so that items may be used. In still others, privileged access to restricted files may be granted to bone fide researchers, however defined by a process (letters of recommendation, IRB approval) that distinguishes them from other people, such as journalists, genealogists, or amateur scholars.

While all users may be told that they should be sensitive to individuals' privacy when they decide to disclose information that they find in the archives and manuscript collections, some are asked to sign confidentiality forms. Others are subject to more stringent legal agreements, worked out in documents designed to give researchers access while giving assurances that no identifiers will be shared with others. No one actually enforces those agreements with criminal or civil penalties, however. If a researcher discloses names and private information culled from restricted items in a talk, blog, or publication, much then depends on who finds out and whether anyone really cares. Researchers may then be reprimanded by their home universities if they went through their

IRB. They may be banned forevermore from the archives or manuscript collections where they worked. In the end, nevertheless, only harmed individuals may actually sue for harms done.

Archivists and curators vary as much as any other set of people when it comes to having positions on how much control we can, and should, have over what information about ourselves is disclosed to others. For those who strongly believe that individuals should give informed consent to all use of personal data (anything linked to their unique identifiers), then strong protections are in order for all information, not just for sensitive topics, and such protections should extend for a period after death. For those who argue that such control is not only completely unrealistic, but that it also values isolation and autonomy over community and shared human experiences, only a few protections on highly personal subjects (such as sexual transgressions) are in order, and should cease at death. There are, and will be, archivists and curators at both extremes and all along the spectrum between them, and they will, usually in consultation with others, be responsible for deciding on policies about acquisitions, levels of processing, and access restrictions.

Archivists are at the gates. They are not the only ones there, of course. Other administrators, budget officers, perhaps even an institution's attorneys, have a role in making decisions that shape who can see what, and when, and how. The more that historians understand archivists' and curators' privacy concerns, however, the better off we will be when it comes to advocating for access, be that formal lobbying at state legislatures and in Congress for improved open access laws or as individuals negotiating for research access to a particular collection. The extent to which historians wish to become privileged researchers, seeking legitimacy through IRBs or application processes that exclude non-researchers, will be an important question for ongoing discussion.

Legal and ethical decision-making does not stop once a historian has that long-awaited box of papers brought out of the archives and into a research space. Historians who have signed user agreements declaring them responsible for legal claims if they invade the privacy of living people are forewarned; others may still learn the hard way if someone is outraged by what has been published about her or his private life and has the energy and resources to launch a lawsuit. When it comes to the dead, however, the potential for legal action against historians vanishes as a reason to be squeamish about sharing details about people with the world. Instead, sensitivities about privacy take over as historians choose what to write and whom to name.

Chapter 5

Managing Privacy

Historians at Work

Chapter 4 should leave no one in any doubt that archivists, curators, and records managers largely control historians' access to the records that document the past. Just as historians share stories about their frustrations with access to information, so too they delight in tales of perseverance that pay off when that key report, that vital bundle of letters, that lost ledger finally appears on the reading room table of the archive or on the out-of-the-way desk at a court house. What historians then do with the names and personal details of those whose lives appear in the records depends a great deal on the questions historians are asking, the professional conventions of those who produced the documents, the kinds of stories they want to tell, and their sense—often hard to articulate—of what is right and wrong. In this chapter, I argue that historians making decisions about identifying their historical subjects have come to know those past people in multiple ways: as individuals constructed within professional or bureaucratic categories (patients, plaintiffs), as victims (of violence, rape), as autonomous beings (doctors, judges). We develop relationships with these constructed people that shape our choices about naming them. As important as our intellectual and emotional connections with our subjects are as we decide to give or to obscure their names, however, historians also need to take care that we do not succumb to the sentimentalities of memory or the valorization of posthumous reputations when we deal with the dead.

Historians think about privacy at three points when dealing with sources. First, historians and their families have been historical actors. They accumulate family letters, personal papers, business records, research notes, professional correspondence, and unpublished works just as other people do. They then have to decide if these materials should be given to an archive for researchers to use. In the process, they have to consider the implications of public access for their own reputations and for family feelings. Second, historians seek access to unpublished records that they need for their work; few historians work entirely from published sources. Sometimes sensitive materials come into their hands without, or in spite of, archival processing, and they must decide how to proceed with using them for research. Third, historians write for publication. There they decide how much to make public: silence? group generalizations supported by technical reference notes to individuals' data? pseudonyms? real names in the text?

Those who spend time thinking about what it means to do history, to engage with the fragmentary sources remaining from the past, to think historically, agree that finding, understanding, and conveying context is everything. "It depends" could be the historian's motto. Thus, this chapter includes a set of stories, some very short, some more lengthy, about how historians have made decisions to protect, or not to protect, the privacy of the people revealed in their research. I make no claim that these stories represent the decision-making processes of all historians, or that they could somehow be used to arrive at inductive generalizations about ethical principles for historical research. On the contrary: context is everything. Nevertheless, the stories do suggest some commonalities. The stories also provide guidance, even wisdom, for historians who may be struggling with worries about what to reveal and whom to name. Ideally, too, they offer non-historians insight into the ways that historians think about their obligations to people and to the past.

Because this chapter is based on what authors have divulged about their choices in their publications and on what a relatively small number of historians have told me in interviews, readers may find one evident bias rather irritating. It has been exceedingly difficult to find reliable examples of historians behaving badly. I asked archivists, I asked historians, I ran uncounted searches. There is gossip and innuendo. Some archivists expressed bitterness with researchers who did not treat living people as well as the archivists thought they should; a few historians thought they had heard something about someone inappropriately revealing names in his or her work.[1] Even under the umbrella of institutionally approved IRB consent forms, in which I promised to mask or anonymize the identity of the interviewee whenever she or he requested it, my informants found it hard to come up with sound instances of what they thought was unethical behavior or cases where they thought that living people have been harmed by the exposure of private details about the living or the dead. It would be nice to believe that they had difficulty because historians actually behave quite well. I suspect, instead, that historians are no more innately ethical than members of other professions.[2] The extent to which people have suffered enough psychological, economic, social, or reputational distress to warrant monitoring historians' research methods, however, remains an unanswered empirical question.

Biographies and Exposés

Before considering how historians have made choices about how to handle privacy concerns when they deal with their sources, we must remember that academic history is part of a huge world of both popular and scholarly publishing. Two examples of books, which coincidentally are both by English professors, underscore the fact that our moral universe as authors is much larger than that set by university IRBs and privacy laws. The first book is a biography. Biography is obviously not history, although biographies frequently convey much rich historical detail and histories often provide intimate portraits of individuals. The former, by definition, center on the individual and help us to understand the complexities of a single life; the latter take a broader view, even if individuals are the focus, and help us to understand the complexities of events and ideas at particular times and places. Biographies need the personal and the private, although older biographies sometimes omitted seemingly irrelevant or unsavory sexual intimacies.[3] The second book is an exposé. Exposés are not history. Although some historians do seek to uncover past wrongdoing, and sometimes construct caricatures of events populated by heroes, villains, and victims, academic historians frown on simplistic accounts and overt judgmentalism.

When published in 1991, Diane Wood Middlebrook's biography of Anne Sexton was immediately controversial. Sexton, a noted poet of the 1960s and 1970s, committed suicide in 1974 and left behind vast quantities of personal papers, drafts of her work, and most intriguingly, some tapes of sessions she had with her psychiatrist, Dr. Martin Orne. Orne later released more session tapes to Middlebrook and she quoted extensively from them in her book. The result is a remarkable study of a creative, troubled woman whose depression was inextricably connected to her poetry, and whose poetry expressed vital aspects of being a woman in the 1950s to early 1970s. Physicians immediately condemned Orne for his decision to let Middlebrook use the session tapes, although he had unequivocal permission from Linda Grey Sexton, Anne's daughter and literary executor.[4] A group of them submitted a formal complaint to the American Psychiatric Association, but nothing came of it because he had Linda Grey Sexton's consent.

Middlebrook reflected on the ethics of her use of the tapes, and much else that was quite intimate in Sexton's materials, including hospital records, in her lovely essay, "Telling Secrets." Linda Grey Sexton gave Middlebrook access to everything her mother kept: "my mother had no sense of privacy," she explained, "and I don't believe it is my place to construct one on her behalf."[5] Middlebrook took pains to say that she did try to respect Linda Grey Sexton's privacy, however, and that of other family members and those who were not already in the public eye, although this reticence did not apply to the names of Sexton's sexual partners. Middlebrook went on to tell an important story in her essay. While living, Sexton had appointed her own biographer, Lois Ames, and when she arranged her papers she included one that she marked "Never to be seen by anybody but Lois Ames. Never to be published." Linda Sexton kept it with Anne Sexton's other papers and, after choosing Middlebrook as biographer instead of Ames, decided that by "Ames" her mother meant "biographer," and so gave Middlebrook access.

Middlebrook had no qualms about opening the file. Her approach to this decision, which violated Sexton's explicit statement, is worth quoting at length:

> I had no intention of obeying the wishes Anne Sexton wrote down, signed and dated in 1973. This is my reasoning: the dead cannot have wishes, they can only have wills, and wills delegate the responsibility for making decisions. I shared Linda Sexton's view that her mother apparently wished to withhold nothing from her biographer. But I do not believe that such a conception of one's subject constitutes an ethical justification—quite the opposite. Claiming to know what the dead would have wanted is usually a self-serving ploy of interested parties, I have found. Thus what Sexton's attitude might have been toward my use of any materials whatsoever—including the therapy tapes—is a meaningless question, in my view, because the dead cannot be asked to make contextual judgments as the living can. And though the dead cannot be consulted, they can also not be shamed or in any way hurt by disclosures of what really happened to them, as the living can.[6]

Middlebrook read the file, only to discover it to be quite anticlimactic. No fascinating revelations; no shocking accounts. It just contained copies of some very early poems that she had sent to her mother, poems she must have felt were not good enough for others to read.[7] What seems embarrassing to us may, indeed, be trivial to others.

The second work also disclosed medical information about deceased individuals, but in a dramatically different context. Martha Stephens published her book, *The Treatment: The Story of Those Who Died in the Cincinnati Radiation Tests*, in 2002. The book is largely a thematic autobiography, not a historical analysis, for it is more about how she came to know what she knew, and what she did about that knowledge, than it is a contextual history of the total body irradiation experiments at the Cincinnati General Hospital between 1960 and 1971. As part of a group of faculty concerned about Department of Defense (DOD) funding for research at the University of Cincinnati during the heightened anti-establishment feelings of the late 1960s and early 1970s, Stephens repeatedly sought copies of research reports sent from university radiologists to the DOD and was, rather surprisingly, given 600 pages of them in the fall of 1971. These served as the basis for an unsuccessful effort to raise a public outcry in Cincinnati against the research project in 1972.

The local activists' report on the cases was sent to Senator Kennedy while he was preparing for Congressional hearings on human experimentation and the Cincinnati research became part of the national coverage of experiments performed without fully informed consent. According to Stephens, during the 1972 Kennedy hearings the University of Cincinnati medical school agreed to stop the experiments as long as Kennedy's staff did not press for the patients' medical and research records and question the Cincinnati physicians before Congress.[8] What incensed Stephens at the time, and for decades afterwards, was the lack of serious press coverage in Cincinnati, the lack of investigations by the University of Cincinnati into the medical school, and, most seriously, the lack of justice shown to the patients, and the patients' families, believed to have been seriously harmed by the experiments.

When national press interest in a wide range of federally funded Cold War radiation experiments revived in 1994, President Clinton established the Advisory Committee on Human Radiation Experiments, which published its 656-page report in 1996. Stephens again became involved in seeking public acknowledgment from the University of Cincinnati that people had been harmed.[9] This time, a volunteer research assistant, Laura Schneider, worked to identify the patients whose initials, ages, and other hints of identifying details appeared in the reports from 1971. Using obituaries, city directories, and other local resources, Schneider succeeded in re-identifying a handful of patients' families. Stephens contacted them, caught the interest of a local human rights attorney, and helped to set in motion the lawsuit that some of the families brought against the medical school and the physicians named in the research documents; publicity about the case helped other patients' families to come forward.[10]

Throughout Stephens's account of her activism, the experiments, and the litigation, Stephens identified deceased patients and their family members by full name wherever possible. She included information from the DOD reports and from patients' medical records released to the families during litigation to describe how the patients suffered. One of the key provisions of the final settlement was a plaque bearing all of the experimental/treatment subjects' names on the hospital's grounds, albeit without much of an explanation for the reason they were being memorialized. Stephens included the full list, with race, age, date of irradiation, dose, date of death, and type of cancer, in an appendix to her book.[11] Stephens's work aptly illustrates that re-identifying individuals whose medical privacy had technically been protected with initials in the Cincinnati physicians' reports is entirely possible with enough corresponding information, such as death dates and ages. The HIPAA Privacy Rule now prohibits the disclosure of such telling bits of data precisely because enterprising sleuths can deduce likely identities and follow up with attempts to contact people or their families.

Stephens expressed only a momentary qualm in her book about her decision to contact the patients' family members—wondering "would the families really want to have this news I would bring them?"—before proceeding to do so.[12] Her concern for social justice impelled her forward. The attorney, Bob Newman, and one of the early family members contacted then made the decision to go public with a press conference to announce their lawsuit. At no point, do note, did Stephens ever describe her efforts as academic research. She and Schneider dug for information like investigative journalists, and to the same ends: public revelations and public accountability. The book she wrote was started after the five years of litigation were well underway.

As an active participant in the events she described, Stephens was deeply invested in criticizing the media, the federal government, the ineffective legal system, the University of Cincinnati, and, most of all, the physicians and other researchers who performed the radiation experiments. For her, the shield of medical authority, played out through claims of expert knowledge—those involved repeatedly asserted that an English professor simply could not understand the benefits of trying high single dose total body irradiation on terminal cancer patients—and the self-serving bulwark of medical privacy allowed doctors to get away with "human sacrifice."[13] Her book, again, is not a work

of historical research. It raises, nevertheless, the stark point that strong protections for the privacy of the records of the deceased protect those in power
just as they protect the vulnerable. Without the ability to identify some of the
patients and to contact their family members, Stephens made it clear that nothing much would have happened beyond passing media interest in 1994. The
media in Cincinnati needed surviving family to describe the very personal stories of their relatives' suffering, to have them become the victims of unethical
experiments. The legal system needed surviving family to engage in a highly
public lawsuit, one that ended with a modest financial settlement for the families and a plaque at the hospital, but no apologies or admission of wrongdoing.

Middlebrook's biography of Sexton and Stephens's exposé of the Cincinnati irradiation experiments to the families of patient participants illustrate the
ethical choices made by two non-historians to include highly personal health
information about recently dead people in their books. They have quite different ends: understanding the complex life of a self-confessed neurotic poet and
seeking recognition for those badly treated by an unapologetic medical system.
Academic historians rarely focus only on the life of a single person and they
even more rarely detail their personal crusades for social justice. Yet some certainly do see their work as contributions to larger discussions of inequalities,
oppression, injustice, and exploitation in American history, and seek the documentation that will reveal the experiences of victims as well as actions of the
powerful. Indeed, some may even choose to share personal and family papers
that future historians could use to study mental illness, abuse, criminality, or a
host of other issues that have shaped our private and public lives.

Donating

Historians with family papers, unpublished professional and personal materials, or private collections of acquired manuscripts at some point have to decide
what to do with them as we age. We may decide some are worth asking the
next generation to save; we may approach an archive with an offer to donate
them; we may just toss them. Perhaps more than non-historians, we may have
a pretty good sense of whether they are worth donating and where they should
go, although curators may politely disagree, and shake their heads over inflated
egos. But, just like non-historians, we have to think about whether our family
papers contain items that are too private, too personal, too embarrassing, or too
alarming to ourselves or to our family members to share with the world.

Thinking about being a donor of personal papers can sharply remind historians about how hard it can be to be generous with messy files and family

intimacies. As Janna Malamud Smith confessed in her own account of planning to destroy her father's papers—and her father was Bernard Malamud, a noted twentieth-century novelist—no matter how valuable they may have been to scholars, they were too personal to share.[14] Constance Putnam, an independent historian-scholar, plans to give her parents' family letters to an archive, but not without some anxiety about how her siblings may feel. Her mother was an avid correspondent and kept round-robin letters going among family members for decades, with her father adding pages when he could. Since her father was a family doctor in a rural village in New Hampshire from the 1930s to the late 1970s, she knows that the letters provide a rich picture of a physician's family life and practice in the mid-twentieth century. Dr. William Putnam rarely revealed the names of his patients or intimate medical details in his letters, but he did tell stories about visiting patients, their foibles, and sometimes his frustrations with them.[15]

What makes Putnam pause as she considers giving the collection to an archive is not the potential revelation of shameful secrets, but her parents' sometimes "disdainful and unkind" comments about rural ignorance and troublesome neighbors. As a historian, she understands the class divide between her parents, with their urban and educated backgrounds, and the farmers and villagers of rural New England. As a daughter, she finds their overt expressions of social and intellectual superiority just a bit embarrassing. The prospect, too, of having to screen between 2,500 and 3,000 handwritten and typed letters for those that might be better off removed from the collection is daunting. She hopes to use them for a book on general practice in mid-twentieth-century New Hampshire—if she lives long enough, as she noted wryly during our conversation. Putnam also acknowledged that she has more concerns about how family members of her generation might view having this material shared than she is about possible feelings of the next generation(s), who barely knew their grandparents or who have moved away from the area.

I, too, have vacillated regularly over the years about what to do with letters and other intimate papers. As a historian, I think someone will find some of them quite interesting someday. As a member of a family, I can only hope that some particularly sensitive items will not cause my brothers' offspring any serious unhappiness. The ones I have no qualms about giving to an archive, if anyone wants them, are the letters my parents and I exchanged when I was a student on my year abroad in England in 1975–1976. I traveled on my own quite a bit and wrote home a lot, with all of the naïve enthusiasms of a college student experiencing British and Continental art, architecture, food, trains, and cheap lodgings. I may even write a note to go with them, confessing where I lied. The other papers are more problematic, although they are far more potentially significant as historical documents about twentieth-century healthcare. To say more would be to disclose too much; I am unable even to construct a plausible hypothetical situation that conveys their potential to harm me personally during my lifetime. Those are the ones I can envision either destroying or closing for a period after my death. What comes first? History or family, especially a future family that is more imagined than real? If I do nothing, will some family member see their value and donate them? Or hastily shred them?

Even historians, in short, pause when facing decisions about letting vague future researchers read and use quite personal papers. Such reticence arises not only for items that could reveal serious misbehavior by ourselves or others, but also for those whose contents are much more mundane. It is hard to grasp a future world where we will not matter as real people to future scholars, where we really will not care if others think badly of us or laugh at our quaint ways. Yet, if we refuse to share, at least to add a postmortem legacy to the historical record, then perhaps we should feel a twinge of hypocrisy.

Researching

The research process is fraught with challenges for historians. Evidence can be so hard to find. Are people who might remember still alive? Can I find them to record their oral histories? Has any documentation that might answer my questions survived? Can I find it? Travel to it? Access it? Ethical concerns are generally far down the list of worries. People will talk, or they won't: they decide. Archives have the documents, or they don't. Archivists have determined they can be used, or they have restricted them. What is left for the historian to ponder? Well, historians do find that even when talking to competent adults, they have to restrain themselves from too much probing. Even when given processed boxes in an archive, historians do come across a misfiled private item and have to decide whether to read it and take notes.

As covered in chapter 2, oral history has attracted quite a bit of discussion about how to do it ethically, especially since it has increasingly come under the purview of Institutional Review Boards (IRBs) because it is research with human subjects.[10] Well before IRBs began to scrutinize oral history, interviewers knew that they had to caution their interviewees not to reveal information that they did not want made public. Susan Reverby recalled a moment when she interviewed a prominent woman in academic nursing who talked about a dispute she had with major figures in her field. "Should I tell you about this?"

the woman asked, and Reverby reminded her that she was going on record. "As a historian," Reverby told me, "did I want to say 'Oh, good, yes, tell me all of it?' Of course. Did I feel like I had an ethical responsibility to her? Yes." After all, Reverby explained, "I went home to history, she went home to the field [she worked in], and she had to live with the published consequences of what she said; I don't."[17] Reverby nicely captured the struggle between curiosity and restraint that oral historians face because their interviewees' audio recordings and transcripts are destined for archival deposit and/or publication. The rigors of IRB-reviewed informed consent procedures may help both interviewer and interviewee review the risks involved in disclosing personal details and dirty laundry. The interviewer must still be ever-mindful that consent is an ongoing process throughout the interview and, as Reverby did, caution the voluble at delicate moments.[18] Once a properly-consented interview recording or transcript is deposited in an archive and opened for use, researchers do not need to have any ethical reservations about using the information it contains.

If archivists were perfect and archival policies unquestionably just, then historians would never have to consider what to do when faced with access to information they are not supposed to have. When Johanna Schoen decided to work on eugenic sterilization in North Carolina for her doctoral dissertation in the early 1990s, she applied for access to the records of the North Carolina Eugenics Board, which spanned the decades from 1929 to 1974. These are held by the state archives, but the state attorney general's office then had control over them. She received access to the board's correspondence and general files, but not to any records about specific sterilization decisions. When she returned to the archive years later, as she worked to turn her dissertation into a book, she asked to see the records again. The archivists asked the state's attorney general for more liberal access on her behalf, and three rolls of microfilm containing detailed minutes of all the board's meetings, including individual petitions and supporting materials, turned up. The archivists refused to copy the microfilm for her, but they did agree to let her print out each page and have the staff cross out all of the names of those to be sterilized. As Schoen studied them at home months later, she observed that internal references indicated that more than one member of a family was sterilized; other internal references showed that more than one petition was sometimes filed for a person when the sterilization order was not granted the first time. The records had complex interconnections. And then she discovered, as others have before her and will again, that if she held the paper up to the light, she could read the names through the marker ink.[19]

The archivists fully intended to protect the privacy of the subjects named in the eugenics records, whether they were living or dead. Schoen needed the names in order to see the connections among families, and not to count the same person twice just because there were two petitions. So, she used them. She created a database for the 7,024 cases, with names. "I have to say that I felt totally justified doing that because for me the ethical question was, how do I write about [the cases]?—not, how do I create the stories? I knew that I would never, ever contact these people or use their names when I wrote about them." It was "the responsible way to do the analysis. I mean, how can I tell the story accurately if I can't do it that way?"[20] In casual conversations over the past five years, I have given perhaps two dozen historians a hypothetical situation based on Schoen's experience, and all immediately have said that they would have made exactly the same choice: gather the data, protect the names.

On May 2, 2002, the governor of Virginia apologized for that state's eugenics program. A reporter at the *Winston-Salem Journal* in December of 2002, having heard about Schoen's work, then wrote a detailed story on the North Carolina's eugenics program, much of it based on Schoen's research. The only subjects contacted for the press coverage were those who had already gone public with their sterilizations because they had launched lawsuits. The governor of North Carolina then apologized for the program and set up the first of several investigative committees to consider compensation for survivors. Ever since Schoen viewed the redacted microfilm pages, the state attorney general has refused all requests for access to the records and has returned control of the eugenics board files to the North Carolina Department of Health and Human Services, although they are still physically housed in the state archive. Individuals can find out if they were sterilized in the program by contacting the Office of Justice for Sterilization Victims.[21] In 2005, well after the initial media and political uproar had died down, Schoen's book, *Choice & Coercion: Birth Control, Sterilization, and Abortion in Public Health and Welfare,* appeared, finally providing the thoughtful historical analysis that the eugenics program so desperately needed, even if cold comfort for its survivors.

Schoen has never revealed the names to anyone. She has been pressed by journalists, researchers, and the sterilized for more information from her files, but has only shared anonymized copies of her database. During our interview in 2011, Schoen discussed her ongoing ethical quandaries. She has an extraordinarily useful database that no one, not even the state archivists, can use because she is not supposed to have the names it contains. She could erase it. Perhaps she should erase it. But something stops her: who else would ever do the work required to extract this data again? Shouldn't it be somewhere safe for future researchers to use after everyone in it is dead? She also has boxes full of poorly redacted copies of archival documents that state officials refuse to let other researchers see. She has thought about having all of her copies scanned as pdf files, which would permanently obliterate the names, sending the scans to North Carolina's state archives for research use, and then destroying all of the copies with readable names. A researcher following IRB requirements to destroy research records containing individually identifiable private information would not pause: into the shredder. Schoen hesitates. Who knows when, and if, the state Department of Health and Human Services will ever release the records for research use? Until she decides what to do, she keeps the data and the documents safe.

Other historians have confided that they have come across very personal items in archival collections, seemingly in places where they did not belong. One, mentioning a private diary discovered among some professional papers, quickly exclaimed: "oh, I shouldn't say that! Don't say where it was!" This researcher did not inform the archivists of the diary's existence: "I was afraid it would be taken away . . . that the material would disappear."[22] Whether this historian's fear was realistic, others share that sense of having found something private out of place, a serendipitous gift of the archival gods. We read these ambiguous items, take notes on them, tuck them back into their folders, and place them back into their boxes to wait for the next historian to come along.[23] Historians have promptly taken some problematic materials, such as documents containing Social Security numbers, to archival staff for reprocessing, of course.[24] Yet, I suspect that a few of us are most helpful about pointing out sensitive information for possible redaction or removal when it has the least potential historical value.

It is impossible to know how often historians face such moments of ethical choice: do I use this data? Do I read the items in this misfiled restricted folder? Should I exploit archivists' mistakes—or what look like mistakes—in processing? What is at stake in making such decisions? From the perspective of the archivist or an ethicist, the historian who continues to read a restricted file is violating the privacy standards established by an institution, whether in accordance with donors' wishes, to comply with privacy laws, or simply as a matter of principle to protect individuals' identities. Historians certainly may respect those standards and still continue to read what they have found because they value the accuracy and completeness of their research. We must no more ignore or suppress evidence available to us than we should make it up. Nevertheless, what feels to historians like dedication to doing the research right might seem to critics to be self-serving excuses for nosiness and professional ambition. Can we distinguish the lure of reading forbidden texts from altruistic devotion to the historian's craft?

We do not have to. As long as the notes and compiled datasets become part of our own protected research files, we can respect the archive's principles, the privacy of possibly living people, and our professional obligations to history. In effect, we redraw the ethical lines set by archivists and records managers and place ourselves inside them. What was confidential stays confidential; the universe of confidants has just increased by one. We compose a mental user agreement with the archive: we will not reveal identities from these restricted sources. We will encrypt the digital files, and keep the documents under lock and key, just as researchers in other fields do with their confidential data. What we learn is then folded into background knowledge, extracted into anonymized datasets, or lost in the noise of evidence culled from other records. If the information is so significant that the historian believes that an individual's identity simply must be disclosed, then it is time to sit down with the archivists (and perhaps the archive's attorneys) to find out how to contact the individual for permission to use the material or to discover if the person has died. Then a review of any agreement for post-mortem closure is in order. Ultimately, the historian will have to decide whether to reveal what he has learned, even in the face of archivists' ire and a living donor's threat of a lawsuit for invasion of privacy.[25]

Writing

Decisions historians make as they face ethically problematic situations while researching are largely hidden ones. Much more obvious are the choices they make when writing, revising, and publishing, since those are the moments when historians move their work from private to public and choose when and how to name people. Some historians do not need to identify people at all. They have data, structural theories, broad historical contexts, and lots of analysis.[26] In many cases, likely even in most cases, the historian decides to use an individual's name when constructing a narrative, a particular story of an event, situation, thought, or relationship. The author ponders composition, not ethics. An analytic passage may need a telling anecdote to provide a metonym for the general argument. The writer hopes that a nuanced account of one episode, one case, or one example will add to an argument's persuasiveness. The historian describes the particular to help readers understand broader relationships among events, trends, cultures, and contexts. The narrative then needs named individuals to move the action along, to pinpoint responsibility for key decisions, or simply because "John Smith" works better than "the officer," or "Jane Smith" better than "the teacher," in the shape of a sentence. At other times "caseworkers," "patient," "doctor," "witnesses," "soldier," or "attorney" may suffice as placeholders for characters the author finds tedious or awkward to list. In this sense, a name may seem to be just another stylistic element to deploy when writing.[27]

Yet, individual names, like other contextual details, also convey authenticity and agency. Just as a historian's credibility as a scholar partly rests on her responsible use of sources and her meticulous citations to them, her credibility as an author partly depends on getting the details right. Accuracy, specificity, and consistency mark truth-telling.[28] When the historian uses a personal name rather than a placeholder such as "officer" or "teacher," he acknowledges uniqueness, even as he includes the name in a story that is supposed to be representative or typical of a larger theme. That uniqueness supports a sense of authenticity because the person named is not a data point, but a real human being, with a place, no matter how small or large, in the past.[29] That place includes making decisions, resisting or submitting to authority, challenging or accepting conditions. To name is to acknowledge historical actors as they shaped as well as when they were shaped by events; as David Gary Shaw put it, for historians, "narrative remains a crucial tool for conceiving and executing agency."[30]

Privacy regulations (e.g. HIPAA Privacy Rule) and archival policies (confidentiality agreements) shape possible narratives because they constrain authors' choices. They apply bureaucratic categories to certain kinds of individuals—victims, patients, juveniles, experimental subjects, vulnerable groups—based on the type of record, not on the actual information included about people. When rules and policies proclaim that individuals' names must not be used, the easiest response is simply not to tell particular stories and to just fold identities into data; the next is to use vague placeholders (patients, victims, delinquents) and to sidestep details that might possibly lead a reader to identify individuals, such as birth and death dates along with occupation, residence, and major life events. So many people are lost to history that the historian essentially preserves the status quo by avoiding names and not making a point of doing so. Another option in some cases is to use a real first name and the initial of the last name. Readers who already know the person's story may recognize the historian's subject, but no one else is likely to do so without a great deal of work. Finally, the historian may use fake names, with or without changing other identifying details, such as occupation or number of children. When a historian replaces a real name with a pseudonym, he performs an interesting slight-of-hand. The name looks authentic, the person with the false name was (perhaps is) real, and the story reads well. Since all of the historians I have found who have used pseudonyms explain that substitution in their footnotes or endnotes, their texts are not even ruffled by the switch. Only meticulous readers discover these small fictions.[31]

Precisely because archivists so often do their jobs well and manage privacy issues before they open collections for research use, historians typically do not hesitate to use the real names that appear in the files they consult in archives and manuscript repositories. Elizabeth Hampsten's 1982 work on the diaries and letters of North Dakota women between 1880 and 1910 offers a prime example. Her book celebrates the private and personal papers that were saved despite the assumption that rural, non-elite women's writing had little historical value. For her, the "phrases 'Read this only to yourself,' 'Don't read aloud,' 'Keep these things to yourself,'" that bracketed parts of the women's correspondence with each other were invitations to reveal their intimate thoughts and experiences of love, sex, reproduction, disease, and death.[32] Distance in time and the high probability that the women were deceased (even if descendants were not) when Hampsten wrote allowed curiosity to overtake any sense of trespass into these women's lives; that, and the fact that others had already explicitly made these private documents available for open historical research.

Legal proceedings provide thousands of documents that are automatically in the public record, albeit one perhaps hard to find if in the basement of an out-of-the-way court house. If a case drew media attention, any private details reported have doubly passed into the public domain. Historians have found legal case files, particularly trial transcripts, to be a rich source of information about the private lives of individuals caught up in criminal cases or civil litigation. Readers of Peter Wallenstein's *Tell the Court I Love My Wife: Race, Marriage, and Law—An American History* (2002) discover story after story of private lives laid bare through the machinations of America's anti-miscegenation laws up to *Loving v. Virginia* in 1967.[33] Janet Golden used many boxes of files on the 1989 *Michael Thorp v. James B. Beam Distilling Company* case for her penultimate chapter in her 2005 book, *Message in a Bottle: The Making of Fetal Alcohol Syndrome.*[34] Four-year-old Michael Thorp and his mother, Candance Thorp, had every facet of their lives, including their medical records, open to probing by Jim Beam's lawyers. Historians, again, have no reason to hesitate to use what such records reveal, for privacy is left at the court room door in the American justice system, unless a judge decides that certain especially intrusive and inflammatory documents should be sealed.

Some historians do pause, nevertheless. Jacki Rand, when working on a book about the history of violence against Native American women in Mississippi during the Jim Crow era, discovered the trial transcript and exhibits used for the prosecution of Alvin Kelly for the rape of a thirteen-year-old Choctaw girl in 1971. The files were saved because Kelly appealed his conviction to the Supreme Court of Mississippi in 1973; the court overturned and Kelly went free. Rand knew that these distressing documents, along with the extensive local newspaper coverage of the case, would be key evidence for her project. Yet Rand felt conflicted about using the girl's name. Rand tracked down the girl's only surviving relative, a cousin, and showed her the transcript. Raking up details from four decades ago would be painful: could she do it? After much discussion, the cousin decided, "I don't want her to be invisible. I want people to know what happened to my cousin. I want people to know her story."[35] At the very start of her project, Rand had also sought the consent of the tribal council of the Mississippi band of the Choctaw Indians for her work in addition to getting her university's IRB approval to interview victims of violence; for her, getting the cousin's consent to use the girl's name, even if not necessary, demonstrated respect for the community she belongs to.

Other historians have been circumspect about naming names from court documents that may be technically part of the public record, but have essentially been made private by obscurity and the challenges of access. Leslie Reagan, for example, extensively used lower-court trial transcripts preserved in the records of the Illinois Supreme Court and Appellate Court, as well as inquest records from the Cook County Medical Examiner's office, in her study of abortion before 1973.[36] In her book, she chose to identify women who were forced to testify against doctors whom the state prosecuted for providing abortions in the 1920s to 1940s using the women's real first names and the initial of their surnames. "I wanted to give these women some protection; they had been forced to testify in order to try to convict someone who had helped them with something that they needed and they were terrified of being in the newspaper."[37] Reagan also decided to do this because some of the women could still have been living. She noted that she would have used full names if she had been certain that the women were deceased, as she did with names in the inquest files. Tracking down death records for the others, however, would have taken more time than she was prepared to invest.

Researchers may obtain documents held by the federal government, but not yet released to the National Archives, through Freedom of Information Act (FOIA) requests. Once released to one person, such documents are essentially considered public, although copies are not necessarily archived for others to use. Starting in the 1980s, Robert Lilly used FOIA requests to access World War II criminal case files held by the U.S. Army Judiciary, Department of the Army, for his work on military justice. Lilly sought trial transcripts and supporting documentation for prosecutions for rape by GIs in the European theater, work that culminated in his *Taken by Force: Rape and American GIs in Europe during World War II* (2007). While he could have used the names of the women whom the Americans assaulted, he chose to identify them, as Reagan did for women who procured abortions, by using only first names and last initials. He chose this method because he "wanted to be as authentic and accurate as possible" without complete disclosure. He always gave the full names of the men convicted of rape whose stories he decided to use. Some of the offenders were dead, because they had been executed for their crimes, but most had been given prison sentences and may still have been living when his manuscript was published. Lilly knew that some Americans would react rather badly to his book and so he wanted to establish an unquestionable evidentiary trail for his analysis.[38]

Susan Reverby decided to suppress the real names of the men involved in the 1932–1972 U.S. Public Health Service's study of untreated syphilis in Tuskegee, Alabama, if their names were not already in publications or other media sources, even though all of the men's medical records are now available without restriction in the National Archives, having been released under FOIA requests. She was especially careful to use pseudonyms when discussing certain details about the men's medical conditions in *Examining Tuskegee: The Infamous Syphilis Study and its Legacy* (2009). She felt that she could make a point that she needed to make—that not all of the men with syphilis actually died of complications from the disease—without publicly contradicting what some of the families believed had happened.[39] The Tuskegee study has aroused too many harsh feelings for yet another white expert to make claims about what is true; if family members want to consult the records, they are free to do so.

All of these authors—Rand, Reagan, Lilly, Reverby—thought hard about naming people, even dead people, who have entered the public record as victims: victims of rape, of criminal proceedings, of exploitation. Such individuals were dragged into the historical record under painful and humiliating circumstances. "I did not want to replicate that [fear of exposure]," Reagan explained.[40]

Protecting these people's full identities was as much about refusing to participate in further victimization as it was about respecting privacy.[41] Reagan stressed that she did use the complete full names of women who had had abortions when she felt that she could: "to not use women's names is to buy into the idea that abortions were shameful." For her to protect the secrets of dead women or to cover up what women freely revealed to others would have been "really problematic in terms of research and in knowing our own history."[42]

What Reagan, Lilly, and Reverby did by choice, many historians of vulnerable populations, such as patients, juvenile delinquents, and welfare cases, have done out of necessity. Hospital records, juvenile court records, welfare records—these all have the privacy of their subjects protected by law or by the policies of the archives that hold them. These laws and policies, in turn, reflect the confidentiality expectations and practices of the professions who deal with the sick, the young, and the surveilled. It is hardly surprising that historians have adopted the standards of the professionals they study, even as they focus on the objects of the professionals' expertise. Well before HIPAA and Privacy Boards, historians of medicine expected that they would have to promise absolute protection for patients' identities if medical records administrators and medical archivists were to allow them to access unredacted patient records. Thus, in the 1990s, when Jack Pressman used the files of the MacLean Psychiatric Hospital for his study of lobotomies in the late 1930s to the early 1950s, he used pseudonyms and carefully noted that he did not even keep a copy of the key that linked these names to case records; it is only available at the archive.[43] Christopher Crenner, a physician-historian, observed that confidentiality was such "a physician thing" that he could not imagine disclosing the names of patients deceased for more than fifty years. "I have adhered to [keeping names confidential] for so long that I would feel myself stepping over a line" even if HIPAA technically no longer protects patients' information at that point.[44]

When Emily Abel used the records of the New York Charity Organization Society for her study of women's responsibilities for caring for sick relatives between 1850 and 1940, she gave pseudonyms to a family whose case from 1918 she recounted, as required by the Community Service Society, the successor group that owns the collection. When I asked her how long identities in such social welfare records need to be anonymized after death, Abel replied, "I think it should be forever . . . it was considered stigmatizing to go to a social service agency. . . . I don't want to add to the insult."[45] For her, even if the people whose lives were probed by early social workers were long dead, they deserved protection.

Obscuring—by choice, requirement, or convention—the names of patients, rape victims, welfare cases, and other unfortunates may protect their privacy if possibly living, or uphold an expectation for posthumous privacy if not. Doing so also reaffirms their status as the recipients of others' actions. Whatever the patients, or victims, or welfare recipients did was reported through the filters of the people with the authority to define what happened.[46] Because experts controlled the documentation of information, their subjects seem to have been powerless. For some historians, to identify them smacks of exploitation, of taking advantage of the vulnerable, the sick, the weak, or the damaged.

At the same time, however, historians frequently name those perceived to have power, whether living or dead. We identify those who wrote the historical sources—the doctors, other health care providers, case workers, attorneys, bureaucrats, business people—because they were obvious historical actors.[47] They made decisions, performed deeds, and did things to other people. In *Deceit and Denial: The Deadly Politics of Industrial Pollution*, one of many possible examples, Gerald Markowitz and David Rosner freely mention industry executives, managers, and chemists who wrote unpublished reports and memos, thus exposing the identities of those culpable of minimizing or covering up the toxic effects of vinyl chloride from the 1930s to the 1970s, a cohort likely to contain the living.[48] These people were hardly public figures; indeed, for those just going about their jobs under the radar of media, they were effectively private citizens who were later caught up in corporate litigation. From the perspective of history, or at least of these two historians, they nevertheless were autonomous individuals who made choices about where to work and how to behave. They were expected to live with the consequences.

Sydney Halpern's research on federally funded human hepatitis experiments from 1942 to 1972 provides an apt example of individuals caught between clear-cut categories of those acted-upon and those who acted. She has dug through partly-processed archives, consulted old newspaper files, and tracked down obscure publications. In the process, she has turned up the names of many experimental subjects. Halpern has no intention of naming the vast majority of them, especially the mentally disabled and prisoners since they are now considered vulnerable populations, unless they are known to be deceased and were mentioned in newspaper stories. Her problem, when I interviewed her in 2012, was what to do with the conscientious objectors during World War II who freely agreed to participate in experiments on hepatitis as an option for alternative service. "The COs weren't just research subjects. They were also historical actors making a statement. They were speaking through their actions.

Historians feel a lot of pressure to withhold names of all research subjects. I know I have. But I think it's a mistake to apply a no-names convention without considering the situation of particular subjects. Leaving COs nameless robs them of a voice in the narrative—it silences them and they wanted to be heard."[49] She plans to use names that are clearly in the public record, as conscientious objectors were individually thanked or acknowledged in various publications; most have been listed on the Civil Public Service website, including those associated with medical research camps, and other authors have routinely named them. At the same time, however, she will suppress identities when discussing information that "could be considered sensitive or embarrassing."

The passage of time will eventually ensure that all World War II survivors are deceased, of course, but the key questions for the writer will remain: were these conscientious objectors vulnerable subjects or historical actors? Do they deserve posthumous privacy or public acknowledgment?[50] We may decide to answer the first question with the statement that conscientious objectors were obviously both subjects and actors; the binary opposition is really not helpful for historical understanding. That careful point does not lead to a way to resolve the binary problem, however. Either these men's real names are used when writing about the experiments, or they are not. What applies to these conscientious objectors can be extended more broadly to other deceased research subjects. According to current ethical standards in human subjects research, they simply must not be named, unless they could be tracked down and confirmed to be deceased, because there are no explicit provisions for releasing names based on a high probability that the subjects are dead, either because they were already quite sick or simply by the passage of time. By definition, by convention, research subjects are anonymized to protect their privacy.[51] Protecting their privacy, in turn, means that they cannot be explicitly recognized as autonomous individuals in the historical events in which they participated. And around we go.

When working on the history of Huntington's disease, Alice Wexler started by planning to use pseudonyms for the eighteenth- and nineteenth-century sufferers of this condition in East Hampton, New York, one of the communities discussed in her book. Since Huntington's is a dominant variation of a single gene, a person with the gene will inevitably develop the disease, and direct familial relationships can disclose the possibility of carrying it. The links between the families she discusses in her book and families affected today have been broken, however, and the unaffected descendants she spoke with "wanted to have the story told." "[P]art of my aim," she wrote, "was to honor the historical presence of these individual lives." In the end, "although medical confidentiality is a critical value, I believe that there is a fine line between protecting confidentiality and perpetuating secrecy and shame." Wexler chose to use the real names of deceased individuals in order to discourage secrecy and to obviate shame, and to place them within the textured lives of their Long Island community. Wexler also chose to discuss her own family's history with Huntington's, including the fact that she learned only when her mother was diagnosed with the disorder that her maternal grandfather had died from it; she grew up with familial and social silences and understood their costs.[52]

Halpern's desire to name her conscientious objectors as historical actors and Wexler's choice to identify sufferers of Huntington's further illustrate a theme running though this section. Historical research is hard work. We spend hours and hours with archival materials and old, obscure publications. We develop relationships with our historical subjects, living and dead. They come to exist in imagined spaces created somewhere between our research notes and our writing. When we write, some we can only see in terms of categories powerfully defined by current professional practices (patients, juveniles); some we voluntarily want to protect, to shield identities abused by others; some we want to ignore; some we want to expose; some we even want to honor. While many historians may make decisions about naming largely in line with historical and professional conventions, abstract rules of research ethics or privacy laws, some find these structures inadequate because they fail to convey the subtleties and richness of historical experiences. These historians see the particular contexts where existing norms, rules, and laws do not make sense. They want to write in unexpected ways, whether that means suppressing names technically in the public domain (witnesses in abortion trials) or revealing names that are usually suppressed (those who participated in medical experiments).

Historians rarely feel conflicted about naming when they write, if only because privacy decisions have long been made by others: the media, the legislatures, the courts, the archivists, and the curators upon whose efforts so much of our knowledge of the past depends. The historians whose decisions I have related in this chapter, however, did have to pause to make self-conscious choices. Except for Christopher Crenner, the physician-historian, none of them spontaneously invoked ethical principles as independent sources of guidance for their work. They referred, instead, to the nuances of specific circumstances and to their own sensibilities about the past. In more abstract terms, when historians suppress the names of the living they are most clearly following the now classical ethical principle of respect for persons, particularly when maintaining privacy around sensitive areas, such as abortion and rape. When historians

suppress the names of the deceased, however, they are following a very different cultural trajectory, one that foregrounds social intuitions about the dead. In
effect, they are valuing memories or—in more technical philosophical terms—
they are acknowledging that the dead themselves may have moral status.

Memory and the Interests of the Dead

Historians have as good a claim as any other group over how to treat the past
in a morally responsible manner—unless they work on events where history
and memory collide. History and memory are very different projects for understanding the past and can demand quite different moral stances. Briefly, and
admittedly far too simplistically, history is the critical academic discipline that
emphasizes the rigorous use of evidence, invites counter-arguments, and seeks
plausible, complex explanations about how and why past events, movements,
concepts, and lives took place and shaped other past events, movements,
concepts, and lives. Memory, in contrast, encompasses efforts to remember,
memorialize, celebrate, mourn, and sometimes, to forgive events, movements,
concepts, and lives of the past.[53] Memory (particularly shared or collective
memories enshrined in traditions and popular history) supports identities,
nationalisms, religions, and ideologies.[54] To remember together is to engage
in a political and ethical process. As Richard Bernstein succinctly put it, "we
feel that certain events and persons ought to be remembered, and we also have
strong views about how they ought to be remembered (or sometimes forgotten)."[55] The "we" is not historians, but anyone invested in past events, from
Holocaust survivors to the families of 9/11 victims.

Those who use oral histories as sources, whether gathered specifically
for a project or consulted in archived transcripts, know well that what people
remember frequently fails to be confirmed in direct primary materials; indeed,
concrete evidence may overtly contradict what a person recollects.[56] A historian treading in these delicate areas may want to respect the testifier and to validate the remembered experience, but he still has to write a narrative faithful to
the entirety of the historical record. Historians can also find themselves protecting embedded family memories, as Susan Reverby did with her decision to use
pseudonyms for some individuals in *Examining Tuskegee* so as not to counter what relatives believed about study participants. Using real names in our
accounts opens very specific past moments to scrutiny by those who remember
(or who remember others remembering) and calls those memories into question. While certainly not as visibly contentious in the United States as claims
and counterclaims over how, when, and where to memorialize certain people

and events, such as Confederate generals of the Civil War, those who examine the histories of regions of recent intense conflict (South Africa, the Middle East) have to carefully manage the multiple realities of witnesses (memory) when they construct synthetic and analytic accounts (history).[57]

Publicly recording the names of the dead is particularly crucial in some memory projects, as long as people are named in order to mourn or to honor them, as the lists carved on war memorials testify.[58] Memorialization, as a community action, preserves public facts as worthy of collective recollection. When we—just ordinary people—imagine being dead, the philosopher Avishai Margalit argues in his *The Ethics of Memory*, we face "the horror of extinction," unless, of course, we hold onto a strong belief in an afterlife where individual identities continue. Failing a religiously-grounded personal afterlife, people can still hope for an insubstantial persistence in the awareness of those who care about them, whether as individuals who sacrificed themselves for their country (war memorials) or simply as a once-living parent, sibling, child, friend, or colleague.[59] We project being remembered by close others, Margalit points out, because that prospect is "an evaluation of the intensity and quality of our . . . relations now, while we are still around."[60]

Except, perhaps, for audacious criminals and despots who hope to achieve a lasting fame by committing horrible acts, people generally hope to be remembered in a positive light, for their good qualities and deeds, not for their mistakes and private indiscretions. We project this sensibility about ourselves as the future dead onto those who are already dead, hence the durability of the ancient phrase *de mortuis nil nisi bene*: do not speak ill of the dead. As Freud remarked in 1915, "this consideration for the dead, which he no longer really needs, is more important to us than the truth and to most of us, certainly, it is more important than consideration for the living."[61] Although "speaking ill" is not the same as "disclosing private information," an impulse to protect posthumous privacy may well draw some of its force from the hope to protect the memory of the dead for the living who remember them.[62]

In his magisterial work, *Posthumous Interests: Legal and Ethical Perspectives* (2008), legal scholar Daniel Sperling takes memories of the dead a step further. He asserts that we must seriously respect the intuitions we hold about the recently deceased: that something of them persists after death for a period of time. He is not asserting a religious or spiritual afterlife for the dead; neither law nor ethics can take such matters of faith into account. He argues, like Margalit, that what persists are our memories of dead individuals, which, among other considerations, means that the dead maintain a "symbolic existence." They are not persons; they have no rights (at least according to law and to most, but not all, theories of rights).[63] But they are not yet totally non-persons. They have interests, he claims, which are the interests of the living that adhere to their postmortem symbolic existences; if these interests are not respected then the symbolic existence is harmed. We must protect against harms to this symbolic existence, including harms to reputation.[64] Even philosophers who disagree with arguments that the dead have interests still assert that if they were to have them, then preserving "a good name" in the memories of the living must be one of them, because it is such a widespread social value.[65] To be clear: it is not the living who are harmed when we weaken the "good name" of the dead in the memories of those who remember them. It is the dead themselves.

Historians who work on sensitive topics involving the recently dead may certainly feel the pull of intuitions about their persistence in the memories of the living, not only by imagining the existence of family members and loved ones who may remember our historical subjects, but also by coming to know the dead through documents about them. The dead can exist for us, too. Danger lurks down this path, however. Fear of distressing imagined relatives suggests an insidious self-censorship for any scholar. If historians set precedents that it is ethically appropriate to protect the reputations of the dead beyond the memories of living relatives and loved ones, we may jeopardize certain historical projects altogether. In a most extreme instance, for example, the philosopher Søren Holm has claimed that it might be ethically wrong to study the DNA of long-deceased historical figures precisely because doing so could harm their existing reputations. Might we not prove that Tutankhamen was the "result of an incestuous dynastic marriage between Akhenaten and his queen Nefertiti" and so "tarnish their good name[s]?"[66] In 2001, Holm advised researchers to consider "whether our scientific question is really important or just represents high level curiosity" before undertaking an analysis of Tutankhamen's DNA. Historical curiosity won out in this case. Analysis of the DNA of eleven Egyptian royal mummies, including Tutankhamen, between 2007 and 2009 revealed considerable consanguinity and lots of health information. The appeal of curiosity—as opposed to "science"— might not be so robust in the future if concerns for genetic privacy expand much beyond protections for the privacy of identifiable living descendants.[67]

Holm is not alone in his worry over the reputations of the dead. Malin Masterton, a bioethicist, and her colleagues have pondered the ethics of possibly discovering through genetic analysis that Queen Christina of Sweden (1632–1654), polymath, intellectual, patroness of Descartes, was actually a man. They suggest that doing so requires serious (scientific) justification. "The interest in

Queen Christiana's gender for historians has dubious reasons. It has been used to discredit the Queen's actions and explain her strong-mindedness, so that no ordinary woman could . . . be the ruler of Sweden."[68] Since discussions of the Queen's purported masculinity have circulated for decades, no doubt determining her chromosomal sex would spark further arguments. That bioethicists do not see the "relevance for historical accounts of her and her life" of this line of inquiry, however, raises red flags. These bioethicists, at least, seem to be valuing memory (reputation) over history (truth-telling) without actually consulting historians.

Tutankhamen and Queen Christina are, quite obviously, long dead. No living person has any direct memory of them, or even a memory of someone who could have known them when alive. They have "reputations" that we remember, then, because they are major named figures that historians (and archeologists, for Tutankhamen) have worked to construct for our schoolbooks, museum exhibits, websites, popular histories, and scholarly publications. Conflating reputations based on memories of the recently living with reputations based on (incomplete, politicized, dated, traditional) constructed historical sources is an obvious flaw in these theorists' arguments.

Conclusion

This chapter offers a handful of examples of the ways that scholars have dealt with privacy and the past. Thinking about these accounts has not led to a grand synthesis, a sweeping manifesto laying out ethical guidelines for historians to follow when hesitating to donate personal papers, when face to face with archives' slip-ups, or when pausing before using a real name in a well-chosen narrative. If anything, this chapter supports the historian's need to make deeply contextual decisions, to weigh the importance of disclosures for a historical argument with reticence about individuals' identities and private details.[69] We must be very careful when succumbing to intuitions about preventing harm to the memories and reputations of the dead, even the recently dead. Unless it is very clear that we are trying to prevent emotional harm to identifiable living people, to protect the names of the dead may only perpetuate stigmas rather than simply acknowledge past prejudices. Historians, too, need to think hard about what it means to respect the dead, if only because some philosophers and ethicists have started to champion protections for the reputations of the deceased that are profoundly ahistorical. I issue a call to resist both overgeneralized ethical positions and inadvertently restrictive regulatory language for research standards to conclude this book.

Chapter 6

Conclusion

Resistance

I believe that Marilyn should have been able to use the real names of the people listed in the nineteenth-century Poor Farm Register from Cedar County, Iowa, in 2006. That county employees applied the 2003 HIPAA Privacy Rule to the register demonstrates the power that a well-intentioned federal regulation had to nearly cripple a research project on nineteenth-century rural poverty. Matthew Warshauer and his graduate student, Michael Sturges, were not so lucky. In 2008, Sturges sought to research the nineteenth-century records of Civil War veterans who had been patients at the Connecticut Hospital for the Insane, now the Connecticut Valley Hospital. They detail their frustrations and determination in "Difficult Hunting: Accessing Connecticut Patient Records to Learn about Post-Traumatic Stress Disorder during the Civil War" (2013). Read their story, especially the remarkable ending: after taking various legal routes for access and actually succeeding in gaining it, the state legislature immediately passed special legislation to keep the records off-limits, although there was a small window of time when Warshauer and Sturges were able to work with them. As Eric T. Dean astutely commented in his reflections on this thematic issue of *Civil War History*: "they are treating the psychological distress of Civil War soldiers and veterans as some kind of awful, deep, dark secret. This is silly."[1] We cannot be the only historians to have run up against the fears that some have about revealing the suffering of past lives to twenty-first-century readers.

This book has explored issues of privacy and the past in research ethics, federal statute and regulatory law, journalism, and tort law. I have delved into archival theory and practice, and have explored how historians make decisions about naming. I sought, and seek, to demystify the abstract notion of protecting privacy by analyzing what regulators, lawyers, jurists, archivists, and historians have said and done in particular instances, although sometimes my actors have been faceless collectives such as Institutional Review Boards (IRBs) and legislators rather than individuals we might hold accountable for their parts in fencing off what we can know about years long gone. I am still bemused by the language of research ethics that encourages IRB members to minimize social and psychological harms that they can imagine happening to research subjects. I much prefer the language of tort law, where only the "reasonable person" can by possibility be truly harmed by "highly objectionable" information being made public. I take heart from the robustness of the First Amendment, although the specter of being sued can cast long shadows.

Privacy is a vital value for the living, at least for the living in the present, in the very recent past, and in certain deeper pockets of time surrounding the intimacies of close relationships, the physical state of our bodies, the mental state of our minds, and the spiritual state of our souls. State and media intrusions into where we live, what we watch, whom we telephone and what we read interfere with our abilities to just be ourselves, with our own opinions, creative impulses, moments of dancing naked to old rock songs, or hours of grieving a beloved pet. Historians certainly should keep their distance from the secrets of the here and now. The secrets of the there and then, when decades have passed, are often less sharp, less embarrassing, and less damaging. At some point, for all of us, they pass into history. If we are concerned about those things that should never be known, then we must by all means reformat the hard drives, shred the letters and diaries, erase the tapes, destroy the photographs, deal with what we may be able to control. For all the shameful items shared with others, we can take heart in realizing how much will never be saved, donated to an archive, seen by a historian, and written about. A good therapist can help with what is left over if something comes to light in such a fashion that people we know can actually identify us. A good lawyer can help when such revelations are seriously misleading and done out of malice.

I cannot leave this book without calling for unqualified resistance to privacy protections for the dead. Well-meaning zeal to build regulatory bulwarks against unwarranted government surveillance, corporate data collection, and industrial-strength network hacking in order to safeguard personal privacy in the present must not be allowed to shape how we can understand the past. We must also discourage those who hope to enhance historians' professional status by articulating ethical statements about respecting the privacy and reputations of the deceased. At the same time, historians should push for stronger research access to records more than thirty years old under agreements to protect the identities of living individuals—but only living individuals—if necessary. Such records then need built-in times when they must be opened to the public, so that genealogists and family historians may freely use them.

We absolutely must resist efforts to extend the language of research ethics in ways that will regulate historical inquiry beyond the (excessive) surveillance already required for oral history. My fear is not that there might be a deliberate, explicit directive to extend research oversight to historical work in public archives or to expand privacy protections for the records of the deceased much beyond what is already covered by the HIPAA Privacy Rule. My fear is that refinements of the language regulating research practices will inadvertently affect historical work precisely because no one is paying attention to the implications that wording for "research" involving "data" has for "data" that were "originally collected for non-research purposes" that "identifies the subjects." Could that include letters or diaries in an archive that might include a few names of the still living? If it could, then inevitably someone will think so, and the Institutional Review Board (IRB) mission creep will creep on. Our vigilance matters so much because I fear that history matters so little to the hundreds of biomedical and social science researchers who do weigh in on the federal regulatory process.

The quotations sprinkled in the previous paragraph are from the advanced notice of proposed rulemaking entitled "Human Subjects Research Protections: Enhancing Protections for Research Subjects and Reducing Burden, Delay and Ambiguity for Investigators" (July 26, 2011). Fortunately, to date this revision to the Common Rule has gone nowhere. The authors suggested that, to reduce the need for IRBs to review data protection expectations for research projects, all research "data" (texts, datasets, images, recordings) that have anything to do with human beings would be covered by a plan to protect privacy. The plan was to be modeled on the HIPAA Privacy Rule for standards of anonymization and security. I learned about this proposed rule when I served on an IRB and it happened to come up in conversation with the staff. It took me days to read it, more days to think I understood it, and many more days to write the critical statement that I submitted to the citizens' comments portal on the regulations .gov website.[2] Oral historians rallied to offer comments, but only a handful of other historians managed to raise concerns about the effects of the proposed new rule on history during the window open for the collection of reactions and opinions. When there is a next time, we will need to do better.[3]

In 2004, *History and Theory* devoted an issue to "Historians and Ethics." It included essays on whether or not historians should offer moral judgments about the past (yes, we cannot help but do so, albeit subtly); whether or not historians have any ethical responsibilities as public intellectuals; how we must counter ethnocentrism; and the like. One author, Antoon De Baets, a Dutch historian who writes in the context of European and international declarations of human rights, proposed "A Declaration of the Responsibilities of Present Generations toward Past Generations." He offered a sweeping set of statements that outline our responsibilities to the dead, where the dead are defined as "former human beings" who deserve "posthumous dignity" because they "retain some traces of human being and personhood after they die."[4] Much of what De Baets lays down are common principles: historians must not lie, deny or omit evidence, distort the truth, or indulge in "unwarranted invasions of privacy." He acknowledges that "the right of historians . . . to know the truth can come into genuine conflict with their duty to respect the privacy and reputations of the dead" and urges us to weigh the benefits and harms. He claims, however, that "bona fide historians respect the dignity of the dead by bringing the past to life but leaving the dead alone."[5]

As admirable as De Baets's impulse is to codify an abstract set of principles for historians to follow, I believe that such efforts are misguided. Where such principles are useful and important, they are redundant. Human beings who ascribe to the various declarations of human rights issued by the United Nations are already obligated to respect the bodies of the dead, for instance, and to reveal human rights abuses.[6] Where the principles are specifically about historians, they are problematic because they assume a universal agreement about the meanings of respect, privacy, and posthumous quasi-personhood. De Baets, so as not to discriminate against the long dead compared with the recently dead, moreover, dismisses intuitions that length of time eases our responsibilities to those who have passed. He advocates universality: all dead of all time are covered by his suggested principles. If we do not accept this, he asserts, we must be able to identify "when 'he' and 'she' become 'it.'" Because that point "would be arbitrary," it would be unjust, especially to the "anonymous dead of history" who are vast in number compared to the "known" dead, which includes the "rich, the powerful, the famous" among those who have been dead for a long time.[7]

What can it mean to respect the privacy of the "anonymous dead of history"? He does not say. I fail to see how oblivion for the anonymous dead conveys respect. Respect for the long-dead rich and famous is misplaced, moreover, if it means suppressing details about their personal lives that are relevant for understanding their historical significance. We must resist the temptation to take a moral high ground by proposing similar ethical statements and hoping that national and international organizations endorse them. History, and the diversity of historical inquiries, is much too complicated for generalized ethical declarations to do us good; they might just as easily cause harm, especially if non-historians were to interpret "unwarranted invasion of privacy" or the sufficiency of our duty to "respect the privacy and reputations of the dead" as a way to censor historical research and writing.

De Baets's call to universalize the way that we treat our dead subjects raises the thorny issue of whether we should even desire to "respect the privacy and reputations of the dead" and, if so, for how long. I repeat: we must resist any efforts whatsoever to enshrine expectations to "respect the privacy and reputations of the dead" in statute law, regulatory law, or declarations of ethical principles. We have left discretion about the dead to donors, archivists, and historians quite well so far. Until there is clear evidence of abuse and harms done to the living, any restrictions based on privacy concerns for the dead must be made by those very familiar with the sensitivity of the information and the specific circumstances of possible disclosure. The degree of vulnerability and exploitation of those named may be an issue for a period of time; extraordinarily sensitive areas involving reproduction, sexual transgressions, illegal acts that could give rise to serious litigation—all of these may or may not require careful handling until enough time has passed.

But how long? Census schedules, the documents that contain information about individuals surveyed for the census, are closed for seventy-two years. Even when those from 1910 were opened in 1982, there must have been living individuals whose household details when quite young were revealed; the most recent to open are those for 1940, and so now include everyone seventy-five and over. The National Library of Medicine closes its collections of patient files for one hundred years after the date of creation if specific death dates are unknown. According to the revised HIPAA Privacy Rule, medical information (but only that held by covered entities) must be protected for fifty years after the death of the identifiable individual, and the death must be documented for any release. So, in late 2018, I could ask Baylor St. Luke's Hospital in Houston to see the medical records of Everett Thomas, the first person to receive a relatively successful heart transplant in the United States. It took place on May 3, 1968 and he lived for 204 days. (I know this because I Googled it; *Life* did a full story on this event, with pictures). Is it likely that the hospital will allow me to research the full details of this historic event and to write about it without restrictions? I suspect not. Yet, if the records of this transplant were simply bundled with those of hundreds of other unidentified people from 1968 and deposited in an archive and we could we apply a one-hundred-years-from-creation rule, at least historians might find them in 2068. More likely they will be destroyed.

Let me take this a step further. If I urged that private information should be closed until the death of the last person who remembers the deceased when living, then I could ask that no one access my maternal grandmother's diary from 1914 until my death, as I remember her when I was three and she was seventy-six, the age at which she died. It isn't a very lengthy memory, but I clearly recall a moment when I saw her coming downstairs in a lacy nightgown. The perception that "the dead [woman] will continue to die long after her physical death, until her memory . . . has also passed into oblivion" may grasp the power of memory, but it would allow me to keep historical documents hostage to suit my own excessive sensitivities.[8] Suppose that such a diary existed in an archive somewhere. How would the archive even know that Helen Crissman's granddaughter is alive? Who would bother to tell the archivists when I die? That burden would freeze historical research on everyone but the childless of the past century. It would also give archivists unbearable headaches. Do not give memory-advocates anything like a claim that we should protect the privacy of the dead until they become the truly dead dead.

I suggest that 115 years from the date of creation be the maximum period of closure for only the most extremely sensitive of records. Records without sensitive information about identifiable individuals are, and should be, opened much sooner than this; ideally, no more than twenty years after the date of creation. Thus, even collections of medical and psychiatric records currently covered by HIPAA would be opened for unrestricted use after 115 years. I urge that medical records belonging to an identified deceased person, moreover, be released twenty-five years after death under a revised HIPAA Privacy Rule. In 2011, only 0.6 percent of the population of the United States was ninety years old or older; far fewer are more than one hundred years old.[9] Given that few medical or other events that occur before age ten are so sensitive that they must be protected, the effective period of closure is 125 years. That is, there might be a very small possibility that the medical records concern a child of ten who lives to be one hundred. If the record was created ninety years ago, twenty-five years have passed when the 115-year limit is reached. The privacy risks to those who are older than one hundred are minimal. Right now, then, we could start to really understand medical practice in 1900, including the connections between hospital patients and their neighborhoods found by tracing names into census records. Just as important, genealogists and family historians could access such records to discover information about their relatives.

Those in favor of long-lasting medical privacy will no doubt laugh at my hubris. Who am I to counter claims that medical confidentiality (the archetype of confidentiality, up there with attorney–client privilege and the sanctity of spiritual confession) has been a cornerstone of medical ethics for hundreds, if not thousands, of years? As I observed in chapter 1, this is not the place to launch a critical historical analysis of this claim, although it sorely needs to be done. I have not found any comments complaining about one of the most massive sets of disclosures of identified health information in American history, for instance, although my searches are still ongoing. The physician-editors of the massive six-volume *Medical and Surgical History of the War of the Rebellion* (Washington, DC: 1870–1888) published thousands of cases of disease and injury among Civil War soldiers, in which they gave full names, ranks, regiments, and companies to most of those included, many of whom survived the war. The cases included injuries to the genitals (with illustrations), which would certainly be considered private by most current standards. Similarly, hundreds of volumes of Civil War hospital registers and thousands of Civil War pension documents—with medical details—have been available to researchers for decades at the National Archives and Records Administration. Has this been a giant ethical blot on the American medical profession? Certainly not. Until there is evidence, based on solid research, that people would refuse to confide in their doctors if they knew that 115-year-old medical records deposited in archives were to be opened to public research, I remain skeptical of the claim that such access would erode trust in the medical profession any more than news reports about doctors' malpractice, fraud, and conflicts of interest already do. Indeed, any set of records more than 115 years old that are protected by professional claims to patient or client privacy are arguably protecting the powerful more than the vulnerable from the scrutiny of historical analysis.

In the end, historians—just like anyone else—must follow their consciences. A historian may need to violate confidentiality agreements if the value of truth-telling or the demands of social justice outweigh privacy concerns, especially if such provisions are largely protecting the interests of powerful people who once engaged in wrongdoing. By the time that a historian is engaged in archival research on a project that has taken months, if not years, of preparation, he is the last person to want to jeopardize his sources by revealing identities without having extraordinarily good reasons. He just has to be prepared to accept the consequences when he does.

At this point, I think it is wise to give Voltaire the last words: "On doit des égards aux vivants; on ne doit aux morts que la vérité." The living deserve our respect; the dead deserve only the truth.[10]

Notes

Chapter 1 — Introduction: The Historians, the County, and the Dead

1. Email communication from Patricia Hamann, Cedar County HIPAA compliance officer, to Marilyn Olson, sent March 23, 2006. Marilyn Olson released this email to me when I was her graduate advisor and has given me permission to use it in this book.
2. Ibid.
3. Marilyn Olson, "'Halt, Blind, Lame, Sick and Lazy': Care of the Poor in Cedar County, Iowa," *The Annals of Iowa* 69 (2010): 131–172.
4. Susan C. Lawrence, "Access Anxiety: HIPAA and Historical Research," *Journal of the History of Medicine and Allied Sciences* 62 (2007): 422–460.
5. Department of Health and Human Services, "45 CFR Parts 160 and 164: Modifications to the HIPAA Privacy, Security, Enforcement and Breach Notification Rules . . . Final Rule," *Federal Register* 78, no. 17 (January 25, 2013): 5613–5614.
6. Jon Wiener, *Historians in Trouble: Plagiarism, Fraud, and Politics in the Ivory Tower* (New York: The New Press, 2005); Ron Robin, *Scandals & Scoundrels: Seven Cases that Shook the Academy* (Berkeley: University of California Press, 2004); see also Thomas Mallon, *Stolen Words: The Classic Book on Plagiarism*, 2nd ed. (San Diego: Harcourt, Inc., 2001).
7. The fundamental question is whether the fiduciary responsibility in the doctor–patient relationship extends to all other relationships and to the record keeper of items after both the doctor and the patient are dead. Ethical codes specify the duty of the physician, but not the duty of all other human beings. Unless doctors take on the responsibility of destroying the records of their own patients, they cannot control the confidentiality of the materials. See the discussion in R. D. Strous, "To Protect or To Publish: Confidentiality and the Fate of the Mentally Ill Victims of Nazi Euthanasia," *Journal of Medical Ethics* 35 (2009): 361–364; strict medical confidentiality has not been observed for the famous: see Philip A. Mackowski, *Diagnosing Giants: Solving the Medical Mysteries of Thirteen Patients Who Changed the World* (New York: Oxford University Press, 2013); nor does it hold in several areas of criminal and civil law as, for instance, in cases of product liability and medical malpractice.
8. Graeme Laurie, *Genetic Privacy: A Challenge to Medico-Legal Norms* (New York: Cambridge University Press, 2002).
9. See the Genetic Information Nondiscrimination Act of 2008.
10. For a particular case study of the effects of the eugenics movement on the lingering shame associated with inherited conditions, see Alice Wexler, *The Woman Who Walked into the Sea: Huntington's and the Making of a Genetic Disease* (New Haven, CT: Yale University Press, 2008), xxi and chaps. 5 and 6.
11. Laurie, *Genetic Privacy*, 94–100, 104–109.
12. Carol Smart, "Family Secrets: Law and Understandings of Openness in Everyday Relationships," *Journal of Social Policy* 34 (2009): 551–567; Anita Vangelisti and John P. Caughlin, "Revealing Family Secrets: The Influence of Topic, Function, and Relationships," *Journal of Social and Personal Relationships* 14 (1997): 679–705;

Evan Imber-Black, *Secrets in Families and Family Therapy* (New York: W. W. Norton, 1993); John Bradshaw, *Family Secrets: What you don't know can hurt you* (London: Piatkus, 1995).

13. Even the National Research Council could find no evidence of social or psychological harm done by the disclosure of research information, although the belief that such harms happen is widespread. Identity theft, for example, appears to result from the theft of data or, in a few instances, from carelessness. The fear that confidential information will be disclosed, either deliberately or inadvertently, is one of the primary reasons why people refuse to participate in surveys. National Research Council, *Expanding Access to Research Data: Reconciling Risks and Opportunities* (Washington, DC: National Academies Press, 2005), 54–56.

14. Chris Dunham, "Genealogy: Another Reason for Your Family to Hate You," entry April 10, 2010, on The Genealogue, a blog, at http://www.genealogue .com/2010_04_01_archive.html (accessed August 13, 2010); Thomas Fiske, "Don't Dig Up the Past!" entry on GenealogyBlog, April 13, 2010, at http://www.genealogyblog .com/?p=8231 (accessed August 13, 2010).

15. Daniel J. Solove, *The Future of Reputation: Gossip, Rumor, and Privacy on the Internet* (New Haven, CT: Yale University Press, 2007); Cameron Anderson and Aiwa Shirako, "Are Individuals' Reputations Related to Their History of Behavior?" *Journal of Personality and Social Psychology* 94 (2008): 320–333.

16. Jessica Berg, "Grave Secrets: Legal and Ethical Analysis of Postmortem Confidentiality," *Connecticut Law Review* 34 (2001): 81–122.

17. Ibid.; Justin Silverman, "The Catsouras Photos: Will a Family's Privacy Interest Impede Press Access?" Posted February 11, 2010, Citizen Media Law Project, at http://www.citmedialaw.org/blog/2010/catsouras-photos-will-familys -privacy-interest-impede-press-access (accessed August 13, 2010); *Catsouras v. California Department of Highway Patrol*, 181 Cal. App. 4th 856 (2010).

18. Janna Malamud Smith, *Private Matters: In Defense of the Personal Life* (Reading, MA: Addison-Wesley, 1997), 3–6, 145–172.

19. The ongoing controversy over access to adoption records aptly illustrates the complexity of moving from a period where information about birth parents was hidden from adoptees to one that is more open, with adoptees arguing that others have no place to make decisions about information so important to them. See E. Wayne Carp, *Family Matters: Secrecy and Disclosure in the History of Adoption* (Cambridge, MA: Harvard University Press, 1998); Paul Sachdev, *Unlocking the Adoption Files* (Lexington, MA: Lexington Books, 1989).

20. Wayne J. Metcalfe and Melvin P. Thatcher, "Serving the Genealogical and Historical Research Communities: An Overview of Records Access and Data Privacy Issues," World Library and Information Congress: 74th IFLA General Conference and Council, August 10–14, 2008, full text posted online at http://archive.ifla.org/IV/ifla74/ papers/117-Metcalfe_Thatcher-en.pdf; Martha Macri and James Sarmento, "Respecting Privacy: Ethical and Pragmatic Considerations," *Language & Communication* 30 (2010): 192–197.

21. This book is limited to the context of U.S. law and practices due to the wide range of legislative and regulatory privacy laws in other countries around the world. See Livia Iacovina and Malcolm Todd, "The Long-Term Preservation of Identifiable Personal Data: A Comparative Archival Perspective on Privacy Regulatory Models in

the European Union, Australia, Canada and the United States," *Archival Science* 7 (2007): 107–127.

Chapter 2 — Research, Privacy, and Federal Regulations

1. Laura Stark, *Behind Closed Doors: IRBs and the Making of Ethical Research* (Chicago, IL: University of Chicago Press, 2012), 81–135.
2. Robert Charrow, "Censorship and Institutional Review Boards: Protection of Human Subjects: Is Expansive Regulation Counter-Productive?" *Northwestern University Law Review* 101 (2007): 707–718.
3. Ezekiel J. Emanuel, Trudo Lemmens, and Carl Elliot, "Should Society Allow Research Ethics to Be Run As For-Profit Enterprises?" *PLOS Medicine* 3 (2006) at http://www.ncbi.nlm.nih.gov/pmc/articles/PMC1518668 (accessed August 8, 2011); Scott Burris and Jen Welsh, "Censorship and Institutional Review Boards: Regulatory Paradox: A Review of Enforcement Letters Issued by the Office for Human Research Protection," *Northwestern University Law Review* 101 (2007): 676.
4. 45 CFR 46.107. For an important analysis of how IRBs actually work, see Stark, *Behind Closed Doors*.
5. The National Research Act (Pub. L. 93–348 signed into law July 12, 1974) created the National Commission for the Protection of Human Subjects of Biomedical and Behavioral Research. The Regulations for the Protection of Human Subjects of Biomedical and Behavioral Research are laid out in 45 CRF 46 subpart A. Subpart B (1974) covers pregnant women and fetuses; subpart C (1978) covers prisoners; subpart D (1983) includes special protections for children. In 1991, sixteen federal agencies brought their regulations in line with 45 CFR 46 subpart A, which is now informally known as The Common Rule. The adoption of the other subparts varies according to agency.
6. Zachary Schrag, *Ethical Imperialism: Institutional Review Boards and the Social Sciences, 1965–2001* (Baltimore, MD: Johns Hopkins University Press, 2010), 78–79; Robert Amdur, *Institutional Review Board Member Handbook* (Sudbury, MA: Jones and Bartlett, 2003), 23.
7. The full text of the *Belmont Report* may be found online on the website of the Office of Human Subjects Research, National Institutes of Health (http://ohsr.od.nih.gov/guidelines/belmont.html). The language of the Common Rule states: "*Research* means a systematic investigation, including research development, testing and evaluation, designed to develop or contribute to generalizable knowledge." 45 CFR 46.102(d) (emphasis in original).
8. One of the best surveys of how historians understand the past is still Peter Novick, *That Noble Dream: The Objectivity Question and the American Historical Profession* (New York: Cambridge University Press, 1988).
9. Zachary Schrag explains why the authors did not modify "research." The Commission did not include any representatives from the humanities, and barely consulted with social scientists, and so conceived of research entirely within a scientific context. *Ethical Imperialism*, 60, 69–70.
10. 45 CFR 46.02(f).
11. Schrag, *Ethical Imperialism*, 154–158; Linda Shopes, "Human Subjects and IRB Review," at http://www.oralhistory.org/do-oral-history/oral-history-and-irb-review. See also, American Association of University Professors, "Regulation of Research

on Human Subjects: Academic Freedom and the Institutional Review Board" (2013) at http://www.aaup.org/report/regulation-research-human-subjects-academic-freedom-and-institutional-review-board

12. 45 CFR 46.101(b)(2); emphasis added.

13. Stark, *Behind Closed Doors*; Charles L. Bosk and Raymond G. DeVries, "Bureaucracies of Mass Deception: Institutional Review Boards and the Ethics of Ethnographic Research," *Annals of the American Academy* 595 (2004): 249–263.

14. 45 CFR 46.102(d). "Oral History and Human Subject Research," UCLA Office of the Human Research Protection Program at http://ohrpp.research.ucla.edu/pages/oral-history-hsr (accessed July 18, 2011).

15. Shopes, "Human Subjects and IRB Review," at http://www.oralhistory.org/do-oral-history/oral-history-and-irb-review. This article, along with Zachary Schrag's book, is a must-read for anyone wishing more information on the complex history and current status of oral history and compliance with the Office of Human Research Protections interpretations of 45 CFR 46.

16. 45 CFR 46.102(d). "Oral History and Human Subject Research," UCLA Office of the Human Research Protection Program at http://ohrpp.research.ucla.edu/pages/oral-history-hsr (accessed July 18, 2011).

17. University of Nebraska–Lincoln, Human Research Protection Policies and Procedures, Policy No. 001, "IRB Review of Oral History Projects," October 1, 2008, at http://research.unl.edu/orr/docs/UNLOralHistoryPolicy.pdf (accessed July 18, 2013); Valerie J. Janesick, *Oral History For the Qualitative Researcher: Choreographing the Story* (New York: Guilford Press, 2010), 146–148.

18. The enabling language for this provision occurs at 45 CFR 46.109(a) and (d).

19. Patricia Cohen, "As Ethics Panels Expand Grip, No Field is Off Limits," *New York Times*, February 28, 2007; see also the reports on IRB reviews and concerns by the American Association of University Professors on their website at www.aaup.org.

20. L. Randall Way, "ORS and IRB Investigation and Punishment of Alleged Faculty Misconduct," *The Faculty Advocate* 8 (2008), n.p. University of Missouri–Kansas City at http://cas.umkc.edu/aaup/facadv23.html#Wray. See Zachary Schrag's blog at http://institutionalreviewblog.blogspot.com/ for reports of cases of reprimand.

21. Oral History Association, www.oralhistory.org, lists publications, resources, and news about the discipline. Also see: Janesick, *Oral History For the Qualitative Researcher*; Barbara W. Sommer and Mary Key Quinlin, *The Oral History Manual* (New York: Altamira Press, 2002), part of the American Association for State and Local History Book Series; the classic Donald A. Ritchie, *Doing Oral History* (New York: Twayne, 1995); John Neuenschwander, *A Guide to Oral History and the Law* (New York: Oxford University Press, 2009).

22. Janesick, *Oral History for the Qualitative Researcher*, 54, notes that "we want to protect all participants by getting IRB approval for any research"; Shopes, "Human Subjects and IRB Review," at http://www.oralhistory.org/do-oral-history/oral-history-and-irb-review.

23. Matt Bradley, "Silenced For Their Own Protection: How the IRB Marginalizes Those It Feigns to Protect," *ACME: An International E-Journal for Critical Geographies* 6 (2007): 339–349.

24. Amdur, *Institutional Review Board Member Handbook*, 28.

25. National Research Council, *Expanding Access to Research Data: Reconciling Risks and Opportunities* (Washington, DC: The National Academies Press, 2005),

51; Laurie Graeme, *Genetic Privacy: A Challenge to Medico-Legal Norms* (New York: Cambridge University Press, 2002), 126–127. The authors of both of these texts assert that revealing identities in and of itself does not support "respect for persons."

26. The *Belmont Report*: "There are, for example, risks of psychological harm, physical harm, legal harm, social harm and economic harm and the corresponding benefits. While the most likely types of harms to research subjects are those of psychological or physical pain or injury, other possible kinds should not be overlooked." For discussions of the lack of evidence that social science research at the time of the *Belmont Report* actually caused any harms, see Edward L. Pattulo, "Institutional Review Boards and Social Research," in *NIH Readings on the Protection of Human Subjects in Behavioral and Social Science Research*, ed. Joan E. Sieber (Frederick, MD: University Publications of America, 1984), and Philip Hamburger, "Censorship and Institutional Review Boards: Getting Permission," *Northwestern University Law Review* 101 (2007): 466–473.

27. The *Belmont Report*: "Risks and benefits of research may affect the individual subjects, the families of the individual subjects, and society at large (or special groups of subjects in society). Previous codes and Federal regulations have required that risks to subjects be outweighed by the sum of both the anticipated benefit to the subject, if any, and the anticipated benefit to society in the form of knowledge to be gained from the research. In balancing these different elements, the risks and benefits affecting the immediate research subject will normally carry special weight. On the other hand, interests other than those of the subject may on some occasions be sufficient by themselves to justify the risks involved in the research, so long as the subjects' rights have been protected. Beneficence thus requires that we protect against risk of harm to subjects and also that we be concerned about the loss of the substantial benefits that might be gained from research."

28. Amdur, *Institutional Review Board Member Handbook*, 86; Richard T. Campbell, "Risk and Harm Issues in Social Science Research," unpublished conference paper for the Human Subjects Policy Conference, University of Urbana–Champaign, April 2003, pp. 3–7. (Available online at www.uiuc.edu\cas\cas_irb\ and downloaded in spring 2004; since withdrawn).

29. 45 CFR 46.102(h)(i); for a critique of minimal risk defined as the discomforts of ordinary life compared with what the law allows as minimal risk for conditions of employment, see Matthew W. Finkin, "Academic Freedom and the Prevention of Harm in Research in the Social Sciences and Humanities," unpublished conference paper, for the Human Subjects Policy Conference, University of Urbana–Champaign, April 2003. (Available online at www.uiuc.edu\cas\cas_irb\ and downloaded in spring 2004; since withdrawn).

30. Bradley, "Silenced for their Own Protection," 339–349. This example is not about an IRB and a historian, but provides an important example of a university IRB prohibiting a social science researcher from making a documentary film with the main content chosen and articulated by the subjects. One concern was that a subject might admit to criminal behavior. For a potent example of the potential minefield of collecting the oral histories of people that included past illegal and violent acts, see coverage of the Belfast Project, for which former IRA and Ulster Volunteer Force members were interviewed about their activities during the Troubles of the 1960s to 1990s. The project did not undergo IRB review; it certainly would have been a

different project if it had. Christine George, "'Whatever You Say, You Say Nothing': Archives and the Belfast Project," (Masters of Science in Information Studies thesis, The University of Texas at Austin, 2012), 1; a full collection of news stories, essays, and publicly released documents about the controversy over the Belfast Project interviews can be found at the Boston College Subpoena News blog at http://bostoncollegesubpoena.wordpress.com (accessed January 31, 2014). For a summary article, see Beth McMurtrie, "Secrets from Belfast: How Boston College's Oral History of the Troubles Fell Victim to An International Murder Investigation," *The Chronicle of Higher Education* 60 (January 31, 2014): A19–A27. It may be possible for a historian to be granted a Certificate of Confidentiality from the National Institutes of Health. These certificates are empowered to protect research information from law enforcement. See http://grants.nih.gov/grants/policy/coc/.

31. Daniel G. Stoddard, "Falling Short of Fundamental Fairness: Why Institutional Review Board Regulations Fail to Provide Procedural Due Process," *Creighton University Law Review* 43 (2010): 1275–1327; the regulations permit an appeals process, but no evidence has been found that any institution has established one.

32. David B. Resnick and Richard R. Sharp, "Protecting Third Parties in Human Subjects Research," *IRB: Ethics and Human Research* 28 (2006): 1–7.

33. Trevor Woodage, "Relative Futility: Limits to Genetic Privacy Protection Because of the Inability to Prevent Disclosure of Genetic Information by Relatives," *Minnesota Law Review* 95 (2010): 682–713.

34. Resnick and Sharp, "Protecting Third Parties," 6–7.

35. This claim rests on negative results from all of the reading I have done in the ethics literature, as well as full text searches on "third-party harm" in *The Oral History Review* and *Oral History* for all available digitized issues.

36. Joan E. Sieber, "Issues Presented by Mandatory Reporting Requirements to Researchers of Child Abuse and Neglect," *Ethics Behavior* 4 (1994): 1–22. Memorandum Turan Odabasi, Associate General Council, to Dr. Mario Scalora, Chair UNL IRB, and Dr. Dan Hoyt, Interim Directory of Research Compliance, RE Nebraska Statutory Provisions Concerning Mandatory Reporting of Child Abuse and Related Statute of Limitations, August 24, 2009, posted at http://research.unl.edu/orr/docs/nebraskachildabusereporting.pdf.

37. For updated summaries of reporting laws for child and elder abuse in each state, along with information about statutes of limitation for abuse crimes, see "The Laws in Your State," on the RAINN: Rape, Abuse & Incest National Network website at http://www.rainn.org/public-policy/laws-in-your-state.

38. Daniel J. Solove, *The Digital Person: Technology and Privacy in the Information Age* (New York: New York University Press, 2004).

39. Institutional Review Boards develop policies on secondary data set use. See, for example, Cornell University's "Secondary Data Analysis Requiring Review," at http://www.irb.cornell.edu/documents/SecondaryAnalysisReview.pdf   (downloaded August 10, 2011).

40. "Chart 5: Does Exemption 45 CFR 46.101(b)(4) (for Existing Data Documents and Specimens) Apply?" prepared by the OHRP at http://www.hhs.gov/ohrp/policy/checklists/decisioncharts.html#5.

41. Data analysts have recently been able to re-identify individuals from data in "anonymized" datasets, however, so the presumed protection offered by deleting certain data fields appears, to some privacy advocates, to be inadequate. See Paul Ohm,

"Broken Promises of Privacy: Responding to the Surprising Failure of Anonymization," *UCLA Law Review* 57 (2009–2010): 1701–1777.

42. See "About Motorola Heritage Services and Archives," in the "History" section of the Motorola Corporation's website at http://www.motorola.com/Consumers/ US-EN/About_Motorola (accessed August 12, 2011); Frank G. Burke, *Research and the Manuscript Tradition* (Landham, MD: The Scarecrow Press, 1997), 261–264.

43. Lindsay M. Howden and Julie A. Meyer, "*Age and Sex Composition: 2010,*" *2010 Census Briefs* (Washington, DC: United States Census Bureau, May 2011), p. 4, table 2 shows that there were 53,364 people in the Unites States over one hundred years of age in a population of 308,745,538, or 1.72 centenarians per 10,000 people (hppt://www.census.gov/prod/cen2010/briefs/c2010br-03.pdf; downloaded July 18, 2011). The probability that any sample less than 1,000 contains a centenarian is thus very low. The probability that any size of sample containing people over 123 years old is virtually zero, since the life span of the oldest human being so far documented is 123 years. See J. M. Robine and M. Allard, "Jeanne Calement: Validation of the Duration of her Life," in *Validation of Exceptional Longevity*. Odense Monographs on Population Aging No. 8 (Odense, Denmark: Odense University Press, 1999; online at the Max Planck Institute for Demographic Research, http://www.demogr. mpg.de/books/odense/6/09.htm; downloaded July 15, 2011).

44. "Guidance on IRB Review of Research Involving Existing Data Sets," The University of Chicago Social & Behavioral Sciences Institutional Review Board (2007) at http://humansubjects.uchicago.edu/sbrirb/publicpolicy.html (accessed August 10, 2011).

45. "FAQ—IRB Website.doc," Institutional Review Board, Sponsored Programs and Research, Winthrop University, at http://www2.winthrop.edu/spar/ FAQ—IRB Website.doc (accessed August 15, 2011).

46. Ronald F. White, "Institutional Review Board Mission Creep: The Common Rule, Social Science, and the Nanny State," *The Independent Review* 11 (2007): 547–564; Frederic L. Coe, "Censorship and Institutional Review Boards: The Costs and Benefits of a Well-Intended Parasite: A Witness and a Reporter on the IRB Phenomenon," *Northwestern University Law Review* 101 (2007): 723–733.

47. For a discussion of the fact that historians cannot claim any special access to information under FOIA for historical research, see *Irons, et al. v. Federal Bureau of Investigation*, 880 F.2d 1446 (1989); Philip H. Melanson, *Secrecy Wars: National Security, Privacy and the Public's Right to Know* (Washington, DC: Brassey's Inc., 2001); Alasdair Roberts, *Blacked Out: Government Secrecy in the Information Age* (New York: Cambridge University Press, 2006); Fred Jerome, *The Einstein File: J. Edgar Hoover's Secret War Against the World's Most Famous Scientist* (New York: St. Martin's Press, 2002).

48. Solove, *The Digital Person*.

49. See sec. 2 of Public Law 93–579 (The Privacy Act) included in 5 USC 552a

50. 5 USC 552a; see *United States of America v. David S. Chase, M.D.*, 2005 U.S. Dist. LEXIS 38676 for an example of the courts upholding the HIPAA clause that gives the Office of the Inspector General access to Medicare billing records for its investigation into Medicare fraud; also *Acara v. Banks*, 470 F.3d 569, 571 (5th Cir 2006); *Logan v Dept of Veterans Affairs*, 357 F. Supp. 2d 149, 155 (D.D.C. 2004).

51. Burke, *Research and the Manuscript Tradition*, 239–244.

52. In addition to the texts of the acts and regulations themselves, see Jennifer A. Mencken, "Supervising Secrecy: Preventing Abuses within Bank Secrecy and Financial Privacy Systems," *Boston College International and Comparative Law Review* 21 (1998), article 5.

53. For a comprehensive discussion of IRBs and litigation, see Sharona Hoffman and Jessica Wilen Berg, "The Suitability of IRB Liability," *University of Pittsburg Law Review* 67 (2005): 365–427. All of the lawsuits brought against IRBs have involved biomedical research and harms under malpractice law, or similar laws; none has challenged the Common Rule as a basis for regulating research itself. Cases naming individual IRB members are rare; at least one was dismissed because the court decided that the federal regulations "create no private cause of action" (384).

54. For HIPAA, see *John P. Cassidy v. Mark Nicolo, Larry Baker, Dr. Gonzalez*, 2005 U.S. Dist. LEXIS 34160; Cassidy sued the defendants for violating his HIPAA rights and the case was dismissed, confirming that HIPAA does not authorize "a private right of action" against individuals. For FERPA, see *Gonzaga University v. John Doe*, 536 U.S. 273 (2002).

55. See, for example, 12 USC 35 (Right to Financial Privacy) §3417 Civil Penalties; 15 USC 41 (Fair Credit Reporting) §1618; 18 USC (Video Privacy) §2707; 47 USC (Telephone Privacy) §227(b)(3).

56. See, for example, the Privacy Rights Clearinghouse, a non-profit organization dedicated to providing information about privacy rights and perceived threats to privacy for consumers at www.privacyrights.org, and the Electronic Privacy Information Center, at www.epic.org.

57. To determine that the privacy laws listed in table 2 do not apply to the records of the deceased, I used the GPOAccess website to open the text of each set of laws in the United States Code and regulations in the Code of Federal Regulations, and searched for "deceased," "decedent," and "posthumous." These are the three words I have found most useful for searching for comments on the privacy of the dead in law review articles and court cases, and hence the likeliest terms to be used in statutory or regulatory law. See, for an example of an explicit comment that the Right to Financial Privacy Act does not apply to the records of the deceased, Office of the Treasury Inspector General for Tax Administration, chap. 400—Investigations, section 60.3.1, in which a "person" is defined. "Persons" do not include "deceased account holders." This document is dated October 1, 2009, and is found at: http://www.treasury.gov/tigta/important_foia_ad_oper.shtml (accessed August 31, 2011).

58. Marjorie Rabe Barritt, "The Appraisal of Personally Identifiable Student Records," *American Archivist* 49 (1986): 267; Mark A. Greene and Christine Weideman, "The Buckley Stops Where? The Ambiguity and Archival Implications of the Family Educational Rights and Privacy Act," in Menzi L. Behrnd-Klodt and Peter J. Wosh, *Privacy and Confidentiality Perspectives: Archivists and Archival Issues* (Chicago, IL: Society of American Archivists, 2005), 181–198. Note that schools, colleges, and universities may have more restrictive policies on the preservation of, and access to, student records than those articulated in FERPA.

59. "Privacy of School Records Laws," Law Library: American Law and Legal Information—State Laws and Statutes at http://law.jrank.org/pages/11819/Privacy-School-Records (accessed July 10, 2011). This site describes the basics about FERPA and includes a table showing state laws and how they intersect with

the federal regulations. See also Sonia Yaco, "Balancing Privacy and Access in School Desegregation Collections: A Case Study," *American Archivist* 73 (2010): 637–668.

60. When I gave a presentation on HIPAA, medical privacy, and history to the Milwaukee Medical Society in 2004, one older gentleman in the audience said that he had recently destroyed his grandfather's and his father's records of their medical practices in rural Wisconsin. He disposed not only of notes on cases, but also their business records and, with them, much about the experiences of managing solo practices in the nineteenth and early twentieth centuries. He did so from the belief that no archive could accept them because of HIPAA.

61. 12 USC 35 §3413. Exemptions also include disclosures required by law enforcement, those required to comply with other statutes, or those required by the courts; Peter Williams, "Is Financial Privacy Preventing Legitimate Research?" *I/S: A Journal of Law & Policy for the Information Society* 5 (2010): 555–567.

62. Board of Governors of the Federal Reserve System, Statistics, and Historical Data, at http://www.federalreserve.gov/econresdata/releases/statisticsdata.htm (accessed August 31, 2011); National Center for Education Statistics, U.S. Department of Education, at http://nces.ed.gov. For databases restricted to "qualified researchers" see http://nces.ed.gov/pubsearch/licenses.asp (accessed August 31, 2011). See also National Research Council, *Expanding Access to Research Data: Reconciling Risks and Opportunities* (Washington, DC: The National Academies Press, 2005) for a discussion of why researchers need access to microdata sets that identify individuals and that can be merged for more complete statistical views of target populations.

63. "Guidance on IRB Review of Research Involving Existing Data Sets," The University of Chicago Social & Behavioral Sciences Institutional Review Board (2007) at http://humansubjects.uchicago.edu/sbrirb/publicpolicy.html (accessed September 2, 2011). See the discussion of IRBs and restricted data sets by the Inter-University Consortium for Political and Social Research at http://www.icpsr.umich .edu/icpsrweb/ICPSR/access/restricted/index.jsp (accessed September 2, 2011).

64. See the discussion in Martin E. Halstuk, "Shielding Private Lives from Prying Eyes: The Escalating Conflict Between Constitutional Privacy and the Accountability Principle of Democracy," *CommLaw Conspectus* 1 (2003): 80–82.

65. To submit a FOIA request to the Department of Justice, for instance, see the instructions on the Department's Office of Information Policy website at http://www.justice.gov/oip/.

66. Matthew D. Bunker, "Takin' Care of Business: Confidentiality under the Business Exemption of the FOIA," *Public Relations Review* 21 (1995): 137–149.

67. 5 USC § 552; *Halpern v. FBI*, 2001 U.S. Dist. LEXIS 24536.

68. *Irons, et al., v. Federal Bureau of Investigation*, 880 F. 2d 1446 (5th Cir. 1989).

69. *Schrecker v. Department of Justice*, 349 F.3d 657 (D.C. Cir. 2003); Michael Tankersley and Marka Peterson, *Public Citizen Litigation Group, Brief of Amici Curiae Public Citizen, Inc., et al. in Favor of Appellant in Schrecker v. Department of Justice*, June 24, 3003.

70. *National Archives and Records Administration v. Favish*, 541 U.S. 157, 163 (2004).

71. Ibid., 166.

72. Ibid., 167.

73. Ibid., 174.

74. Martin E. Halstuk, "When is Invasion of Privacy Unwarranted Under FOIA? An Analysis of the Supreme Court's 'Sufficient Reason' and 'Presumption of Legitimacy' Standards," *Journal of Law and Public Policy* 16 (2005): 361–399.

75. Ibid., 392–394.

76. Ibid., 363, 380–383, for a discussion of previous cases where information about the mode or manner of death—almost always in pictures—was withheld to protect the privacy interests of relatives because such a release would be an "unwarranted" invasion of privacy.

77. Searches in *LexisNexis* (all federal and state cases) and *America's News*, a subscription service to national, local, and regional printed and online newspapers, newswires, blogs, and broadcast transcriptions turned up no references to any lawsuits challenging 45 CFR 46 by researchers. There are plenty of lawsuits against biomedical researchers about violations of 45 CFR 46, needless to say.

78. See also: Matthew Warshauer and Michael Sturges, "Difficult Hunting: Accessing Connecticut Patient Records of Learn About Post-Traumatic Stress Disorder during the Civil War." *Civil War History* 59 (2013): 419–452.

79. 45 CFR 160.103; see Susan C. Lawrence, "Access Anxiety: HIPAA and Historical Research," *Journal of the History of Medicine and Allied Sciences* 62 (2007): 422–460; R. D. Strous, "To Protect or To Publish: Confidentiality and the Fate of the Mentally Ill Victims of Nazi Euthanasia," *Journal of Medical Ethics* 35 (2009): 361–364.

80. *State of Nebraska, ex rel. Adams County Historical Society vs. Nancy Kinyoun*, Petition for Writ of Mandamus In the District Court of Adams County, Case No. CI 07–538, February 14, 2008.

81. Ibid. The court had previously required Ms. Kinyoun to produce the records or show cause why she (for the DHHS/Hastings Regional Center) would not. The court heard the arguments for cause on December 6, 2007 and agreed that she did not have to release them.

82. *State of Nebraska, ex rel. Adams County Historical Society vs. Nancy Kinyoun*, Appellant's Reply Brief and Brief in Opposition to Cross Appeal, Case No. A-08-339, October 6, 2008; 45 CFR 164.512. The other state cases are *Cincinnati Enquirer v. Daniels*, 844 N.E. 2d 1181 (Ohio 2006) and *Abbott v. Texas Dep't of Mental Health*, 212 S.W. 648 (Tex. App.–Austin 2006, no pet.).

83. *State of Nebraska, ex rel. Adams County Historical Society vs. Nancy Kinyoun*, Appellant's Reply Brief and Brief in Opposition to Cross Appeal, Case No. A-08-339, October 6, 2008.

84. *State of Nebraska, ex rel. Adams County Historical Society vs. Nancy Kinyoun*, 277 Neb. 749 (2009).

85. The oral arguments heard at the district court hearing have been deduced from material in the appeals' briefs. For the oral arguments before the Nebraska Supreme Court, go to http://www.supremecourt.ne.gov/oral-arguments/sc/archive/3-09-sc-oral-arguments.shtml, and listen to the recordings for S-08-0339 (dcr_20090303_08-0339.wav).

86. The record is now available on the Nebraska State Historical Society website at http://www.nebraskahistory.org/databases/hastings.htm and on the Adams County Historical Society website at http://www.adamshistory.org/.

87. Robert Townsend, "Oral History and Review Boards: Little Gain and More Pain," *AHA Perspectives* (2006), at http://www.historians.org/perspectives/issues/2006/0602/0602new1.cfm.

Chapter 3 — Historians, the First Amendment, and Invasion of Privacy

1. Committing actual research harms are, of course, subject to punishment through the law. In biomedical research, physicians have been sued for malpractice that occurred in clinical trials. See Sharona Hoffman and Jessica Wilen Berg, "The Suitability of IRB Liability," *University of Pittsburg Law Review* 67 (2005): 365–427. Hoffman and Berg outline the possibility for suing IRBs themselves under the torts of negligence, invasion of privacy, breach of confidentiality, and breach of fiduciary duty. As far as I can determine, when such suits have been brought, they have been dismissed without appeal or have been settled out of court.

2. Mark Feldstein, "Kissing Cousins: Journalism and Oral History," *Oral History Review* 31 (2004): 5.

3. American Association of University Professors, "Statement of Principles on Academic Freedom and Tenure (1940)" available at http://www.aaup.org/AAUP/pubsres/policydocs/contents/1940statement.htm; American Association of University Professors, "Regulation of Research on Human Subjects: Academic Freedom and the Institutional Review Board," (2013) at http://www.aaup.org/report/regulation-research-human-subjects-academic-freedom-and-institutional-review-board.

4. One rationale for government interest in regulating research is the precedent established in First Amendment cases that the government has the authority to regulate how information is acquired. Forbidding the use of information acquired from illegal wiretapping, for instance, is allowable. So, too, is forbidding the dissemination of information gained from the discovery process in legal cases if the documents have been put under a protective order. Restraining information gained from these methods does not violate the First Amendment because they do not forbid the publication of the same information if obtained in other ways. See *Seattle Times Co. v. Rhinehart, et al.*, 467 U.S. 20 (1984).

5. James Lingren, Dennis Murashko, and Matthew R. Ford, eds., *Northwestern University Law Review: Special Issue: Symposium on Censorship and Institutional Review Boards* 101 (2007).

6. For up-to-date information on lawsuits involving IRBs and social scientists, consult Zachary Schrag's blog, *Institutional Review Blog: News and Commentary about Institutional Review Board Oversight of the Humanities and Social Sciences,* at http://www.institutionalreviewblog.com/ (accessed June 22, 2012).

7. For a story about researchers who did get into serious trouble with their institution's IRB over an essay that critically analyzed another scientist's work on the legitimacy of recovered memories of child abuse and, in the process, interviewed the anonymous victim's mother and several others involved with the case, see Carol Tauris, "The High Cost of Skepticism," *The Skeptical Inquirer* 26 (2002) online at http://www.csicop.org/si/show/high_cost_of_skepticism; for a description of their study, see Elizabeth F. Loftus and Melvin J. Guyer, "Who Abused Jane Doe? The Hazards of the Single Case History, Part I and Part II," *The Skeptical Inquirer* 26 (2002): 24–32, 37–42.

8. Isabel Awad, "Journalists and Their Sources: Lessons from Anthropology," *Journalism Studies* 7 (2006): 922.

9. Ibid., 928.

10. Ibid., 932.
11. Joseph J. Hemmer, *The Supreme Court and the First Amendment* (New York: Prae-
    ger, 1986), 194; Samuel D. Warren and Louis D. Brandeis, "The Right to Privacy,"
    Harvard Law Review 4 (1890): 193–200.
12. For a sophisticated discussion of the theoretical and empirical status of "freedom of
    the press," see Timothy E. Cook, *Freeing the Presses: The First Amendment in Action*
    (Baton Rouge: Louisiana State University Press, 2005), especially Cook's introduc-
    tory essay, pp. 1–26 and part II; Hemmer, *The Supreme Court and the First Amend-
    ment*, 292–293; Martin E. Halstuk, "Shielding Private Lives from Prying Eyes: The
    Escalating Conflict Between Constitutional Privacy and the Accountability Principle
    of Democracy," *CommLaw Conspectus* 11 (2003): 89–90.
13. See, among many other publications: John L. Hulteng, *The Messenger's Motives: Eth-
    ical Problems of the News Media* (Englewood Cliffs, NJ: Prentice-Hall, Inc., 1976);
    H. Eugene Goodwin, *Groping for Ethics in Journalism* (Ames: Iowa State University
    Press, 1983); Philip Meyer, *Ethical Journalism: A Guide for Students, Practitioners
    and Consumers* (New York: Longmans, 1987); Stephen J. A. Ward, *The Invention of
    Journalism Ethics: The Path to Objectivity and Beyond* (Montreal: McGill-Queen's
    University Press, 2006); Dale Jacquette, *Journalistic Ethics: Moral Responsibility
    in the Media* (Upper Saddle River, NJ: Pearson, 2007); Gene Foreman, *The Ethical
    Journalist: Making Responsible Decisions in the Pursuit of News* (Chichester, UK:
    Wiley-Blackwell, 2010); Christopher Meyers, ed. *Journalism Ethics: A Philosophical
    Approach* (Oxford: Oxford University Press, 2010).
14. Deni Elliott and David Ozar, "An Explanation and a Method for the Ethics of Jour-
    nalism," in Meyers, *Journalism Ethics*, 19–24; Foreman, *The Ethical Journalist*, 83–
    103; "Ethics Code: Associated Press Managing Editors," Appendix 3 in Jacquette,
    Journalistic Ethics, 287–288; "*New York Times* Company Policy on Ethics in Jour-
    nalism," The *New York Times* Company, at www.nytco.com/press/ethics.html
    (accessed June 19, 2012).
15. Elliot and Ozar in Meyers, *Journalism Ethics*, 19–24; see also Jacquette, *Journalistic
    Ethics*, 126–154, 178–207.
16. Lee C. Bollinger, *Uninhibited, Robust, and Wide-Open: A Free Press for a New Cen-
    tury* (New York: Oxford University Press, 2010); Cook, *Freeing the Presses*; Clifford
    G. Christians, "The Ethics of Privacy," in Meyers, *Journalism Ethics*, 224. Christians,
    a media studies scholar, argues that respect for persons should trump easy assertions
    of "the public's right to know." Aaron Quinn, "Respecting Sources' Confidentiality:
    Critical but Not Absolute," in Meyers, *Journalism Ethics*, 272–273.
17. Robert D. Richards and Clay Calvert, "Suing the Media, Supporting the First
    Amendment: The Paradox of Neville Johnson and the Battle for Privacy," *Albany
    Law Review* 67 (2004): 1097–1135.
18. Again, it is important to stress that this question applies to historians in the United
    States. It is much easier for people to sue historians for defamation in Europe and
    elsewhere. See Antoon De Baets, "Defamation Cases Against Historians," *History
    and Theory* 41 (2002): 346–366.
19. All of the information in this section, unless otherwise noted, derives from Wil-
    liam Prosser and Page Keeton, *Handbook of the Law of Torts*, 5th ed. (St. Paul, MN:
    Thomson/West, 1984, with pocket updates to 1988), chap. 20: Privacy, 849–869.
20. Ibid., 856–857.

21. For recent cases that review this issue, see *Chapman v. Journal Concepts, Inc.*, 528 F. Supp. 2d 1081 (D.Haw. 2007); *Gates v. Discovery Communications*, 101 P.3d 552 (Calif. 2004).

22. Amy Gajda, "Judging Journalism: The Turn towards Privacy and Judicial Regulation of the Press," *California Law Review* 97 (2009): 1057–1060. For a recent case that reviews the challenges of succeeding in a suit against the media for disclosure of private facts (in this case, identities), see *Alvarado, et al. v. KOB-TV*, 493 F.3d 1210 (10th Cir. 2007).

23. Richards and Calvert, "Suing the Media, Supporting the First Amendment," 1133; Hemmer, *The Supreme Court and the First Amendment*, 210–217; *Taus v. Loftus*, 151 P.3d 1185 (Calif. 2007).

24. *Second Restatement of Torts* (1977), quoted in Gajda, "Judging Journalism," 1066; *Taus v. Loftus*, 151 P.3d 1185 (Calif. 2007).

25. Gajda, "Judging Journalism," 1072–1083; see also William Coté and Roger Simpson, eds., *Covering Violence: A Guide to Ethical Reporting About Victims and Trauma* (New York: Columbia University Press, 2000).

26. Gajda, "Judging Journalism," 1093–1104; Eugene Volokh, "Freedom of Speech and Information Privacy: The Troubling Implications of a Right to Stop People from Speaking About You," *Stanford Law Review* 52 (2000): 1049–1124.

27. *Briscoe v. Reader's Digest*, 483 P.2d 34, 36 (Calif. 1971), overruled by *Gates v. Discovery Communications*, 101 P.3d 552 (Calif. 2004).

28. *Roshto v. Hebert, et al.*, 439 S0.2d 428, 431 (Louis. 1983).

29. *Roshto*, 439 S0.2d at 432.

30. See the review of key decisions in *Gates v. Discovery Communications*, 101 P.3d 552 (Calif. 2004); see also *Meeropol v. Nizer*, 417 F Supp. 1201 (S.D. New York 1976), where the author of *The Implosion Conspiracy* was sued by Julius and Ethel Rosenberg's children for invasion of privacy and copyright infringement. All actions were dismissed.

31. *Haynes v. Alfred A. Knopf, Inc. and Nicolas Lemann*, 8 F.3d 1222 (7th Cir. 1993).

32. Ibid., 1227–1228.

33. Ibid., 1231.

34. Ibid., 1232, with parentheses around the second sentence omitted. Nothing in the court's opinion explicitly addresses the question of consent. It is presumed that Lemann had Daniels's consent to use information from her interview, as that is standard journalistic practice. By the time of the book's publication, Daniels was dead, so her perspective on the book, and her ex-husband's decision to sue, is unknown.

35. Ibid., 1224.

36. Ibid., 1233.

37. Ibid.

38. *Gilbert v. Medical Economics*, 665 F. 2d 305, 306 (10th Cir. 1981).

39. Ibid., 308. Justice Posner quoted the second sentence.

40. *Haynes*, 8 F.3d, 1233.

41. Ibid., 1235.

42. Ibid., 1223.

43. Gajda, "Judging Journalism," 1070; these cases were sent back for trial, but the outcomes are not clear. In the one involving NBC's *To Catch A Predator* show, the man's sister, who launched the original suit, settled out of court. Matea Gold,

"NBC Resolves Lawsuit over 'To Catch a Predator,'" *LA Times*, June 24, 2008 at http://latimesblogs.latimes.com/showtracker/2008/06/nbc-resolves-la.html.

44. Eleanor L. Grossman, J.D., "Privacy," *American Jurisprudence*, 2nd ed. (Eagan, MN: Thomson Reuters, 2012), 62 A §14; *Butcher v. The Lincoln Journal*, 2012 U.S. Dist. LEXIS 3883 (2012).

45. *Justice v. Belo Broadcasting Corporation*, 472 F. Supp. 145 (N.D. Tex. 1979); *Young v. That Was The Week That Was*, 312 F. Supp. 1337 (N.D. Ohio 1969).

46. *Catsouras v. Department of California Highway Patrol*, 181 Cal. App. 4th 856 (2010). In this case, members of the Highway Patrol distributed photographs of the death of an eighteen-year-old girl in a car accident by email, and the photographs ended up on the internet, so there was no news context for them at all. In contrast, in *Savala v. Freedom Communications*, 2006 Cal. App. Unpub. LEXIS 5609 (2006), found that the publication of a photograph of a brother's murdered body was newsworthy. Note that this opinion, while available through LexisNexis Academic, cannot be used as a proper citation in a court because "this opinion has not been certified for publication." A mother did successfully sue a publisher for disseminating nude photographs of her dead daughter taken twenty years previously. The grounds were not "public disclosure of private facts," however, but appropriation of her daughter's images in order to gain profit.

47. *Lee v. Weston*, 402 N.E.2d 23 (Ind. App. 1980); Hannes Rosler, "Dignitarian Posthumous Personality Rights—An Analysis of U.S. and German Constitutional and Tort Law," *Berkeley Journal of International Law 26* (2008): 153–205.

48. Raymond Iraymi, "Give the Dead Their Day in Court: Implying a Private Cause of Action for Defamation of the Dead from Criminal Libel Statutes," *Fordham Intellectual Property, Media & Entertainment Law Journal* 9 (1999): 1083.

49. Ibid.

50. Ibid., 1104.

51. Ibid., 1109–1111.

52. William H. Binder, "Publicity Rights and Defamation of the Deceased: Resurrection or R.I.P.," *DePaul-LCA Journal of Art and Entertainment Law* 12 (2002): 297–316; Lisa Brown, "Dead But Not Forgotten: Proposals for Imposing Liability for Defamation of the Dead," *Texas Law Review* 67 (1989): 1525–1567.

53. For a complete review of the legal status of professional relationships and confidentiality, see Ronald L. Goldfarb, *In Confidence: When to Protect Secrecy and When to Require Disclosure* (New Haven, CT: Yale University Press, 2009).

54. The Citizen Media Law Project maintains a website on which cases are posted by topic. "Publication of Private Facts" is one of these, and it is a useful source for ongoing information about cases being filed: at http://www.citmedialaw.org/subject-area/publication-private-facts.

Chapter 4 — Archivists at the Gates

1. The awkward descriptive phrase used here—"material information-containing items"—introduces the slippery problem of just what it is that archives contain, or should contain, compared with what people have put into museums. We now recognize that a great deal more of the past comes to us through artifacts that may or may not have textual inscriptions, including two-dimensional ones such as maps, drawings, photographs, and other images, and three-dimensional ones, such as

film reels, videotapes, computer hard drives, coins, clay tablets, knotted strings, and textiles. Many museums contain documents; some archives contain objects. For the purposes of convenience, "records," "texts," "documents," and "materials" are used in this chapter to refer to material information-containing items, including the physical repositories of electronic information, with clear awareness that doing so is philosophically problematic.

2. Terry Cook, "The Archive(s) is a Foreign Country: Historians, Archivists, and the Changing Archival Landscape," *The American Archivist* 74 (2011): 601.

3. Luciana Duranti, *Diplomatics: New Uses for an Old Science* (Lanham, MD: Society of American Archivists, Association of Canadian Archivists, The Scarecrow Press, 1998), 37; Jacque Derrida, *Archive Fever: A Freudian Interpretation* (Chicago, IL: University of Chicago Press, 1996). The professional literature on archives and archival theory is extensive. Historians new to thinking about archives as a phenomena and a process would do well to start with Francis X. Blouin, Jr. and William G. Rosenberg, *Processing the Past: Contesting Authority in History and the Archives* (New York: Oxford University Press, 2011); Cook, "The Archive(s) is a Foreign Country," 600–632; John C. Rule and Ben S. Trotter, *A World of Paper: Louis XIV, Colbert de Tracy and the Rise of the Information State* (Montreal: McGill-Queen's University Press, 2014).

4. Heather MacNeil, *Without Consent: The Ethics of Disclosing Personal Information in Public Archives* (Metuchen, NJ: The Scarecrow Press and the Society of American Archivists, 1992), 109–116, 167, on the growth of individual dossiers and social history, and the increasing attention that archivists should pay to informed consent and individual autonomy; Amy Fitch (The Rockefeller Archives), telephone interview, July 1, 2014; Stephen Novak (Augustus C. Long Health Sciences Library, Columbia University), telephone interview, May 21, 2014; Judith Wiener (Medical Heritage Center, The Ohio State University), interview, Columbus, OH, June 16, 2014; Chris Paton (Archivist, Columbia Theological Seminary; former archivist at the Georgia Division of Archives and History), telephone interview, July 16, 2014.

5. For in-depth analyses of various case studies, see Menzi L. Behrnd-Klodt and Peter J. Wosh, *Privacy and Confidentiality Perspectives: Archivists and Archival Issues* (Chicago, IL: Society of American Archivists, 2005).

6. Novak, interview, May 21, 2014; Novak noted that donors may die, thus releasing their records, and the archive is not necessarily informed.

7. John D. Wrathall, "Provenance as Text: Reading the Silences around Sexuality in Manuscript Collections," *The Journal of American History* 79 (1992): 156–178; this book does not address the ethical issues raised by historians who use unpublished materials in private collections or that have somehow come into their own possession. Knowingly relying on evidence that is not, and will not be, accessible to other researchers is problematic, if only because it means that no one else can check the accuracy (or even existence) of the information.

8. See, for example, Helena Pohlandt-McCormick, "In Good Hands: Researching the 1976 Soweto Uprising in the State Archives of South Africa," in *Archive Stories: Facts, Fictions and the Writing of History,* ed. Antoinette Burton (Durham, SC: Duke University Press, 2005), 299–324; International Council on Archives, *Memory of the World at Risk: Archives Destroyed, Archives Reconstituted,* Archivum, vol. XLII (Munich: K. G. Saur, 1996).

9. Trevor Livelton, *Archival Theory, Records and the Public* (Lanham, MD: The Society of American Archivists and the Scarecrow Press, 1996); Frank G. Burke, *Research and the Manuscript Tradition* (Lanham, MD: The Society of American Archivists and the Scarecrow Press, 1997).

10. Frank Boles and Mark Greene, "Et Tu Schellenberg? Thoughts on the Dagger of American Appraisal Theory," *The American Archivist* 59 (1996): 301, uses the example of the Constitution to emphasize the symbolic function of the display of archived documents.

11. The Society of American Archivists, the largest association of the profession, has both archivists and curators as members.

12. For a much more detailed and nuanced discussion of "truth" versus "authenticity," see Duranti, *Diplomatics: New Uses for an Old Science*, 45–58; Elena Danielson, *The Ethical Archivist* (Chicago, IL: Society of American Archivists, 2010), chap. 7, "Authenticity and Forgery," 219–246.

13. See the website of ARMA International (formerly the Association of Records Managers and Administrators) for an introduction to the profession of records management: www.arma.org. The association began in 1955. For an example of a handbook for a working institution, see the Thomas Jefferson National Accelerator Facility, Records Management Handbook (n.d.) at https://www.jlab.org/div_dept/cio/IR/records/handbook/index.html (accessed February 11, 2014).

14. "So You Want to Be an Archivist: An Overview of the Archives Profession" at http://www2.archivists.org/profession (accessed February 7, 2014). See also Francis X. Blouin, Jr., "The Evolution of the Archival Practice and the History-Archival Divide," in *Controlling the Past*, ed. Terry Cook (Chicago, IL: Society of American Archivists, 2011), 325.

15. For the extremely complex set of policies directing federal records management and NARA, see "Records Management" on the NARA website at www.archives.gov/records-mgmt/ (accessed February 10, 2014); most agencies have, or will have, management plans in place that allow certain categories of paperwork to be shredded/deleted immediately, after three years, after ten years, and so on. The Federation of American Scientists keeps track of policies and procedures surrounding government secrecy at "Project on Government Secrecy," at www.fas.org/ssp/govsec/index.html (accessed February 10, 2014). This is a useful site for discussions about declassification as an ongoing problem and challenge for NARA. For records management processes at other institutions and for businesses, a basic internet search on "records disposition decision" will produce a range of URLs to pages and documents laying out policies. In general, all require that confidential documents, personnel records, and copies of vital records (birth certificates, marriage licenses) be shredded or otherwise destroyed. For a now-outdated discussion of the wide range of relationships between agency records managers and state archivists, see Roland M. Bauman, "The Administration of Access to Confidential Records in State Archives: Common Practices and the Need for a Model Law," *The American Archivist* 49 (1986): 349–369.

16. See, for example, "New York State Archives Policy on Access to Records," at http://www.archives.nysed.gov/a/research/res_access.shtml (accessed March 13, 2014).

17. The internet as a source repository is discussed in several of the essays in Clare Bond Potter and Renee C. Romano, eds., *Doing Recent History* (Athens, GA: The University of Georgia Press, 2012).

18. The Internet Archive is a massive effort to make digital works freely available to all users, including archived web pages. See "About the Internet Archive" at https://archive.org/about and the "Wayback Machine" at https://archive.org/web/web.php. Archivists and curators are increasingly capturing webpages at specific points in time in order to document significant events. These collections are then stored in repositories such as Archive-It (https://archive-it.org/; Archive-It is a subscription service affiliated with the Internet Archive.) Such collections are properly provided with metadata, and are included as items with entries in cataloging systems. See, for example, the inventory of the Matthew Shepard Web Archive, 1998–2008, created by the American Heritage Center, University of Wyoming, at http://rmoa.unm.edu/docviewer.php?docId=wyu-ah300023.xml. The pages are hosted on Archive-It. I am grateful to Mark Greene for this example, and for his point that archivists and curators are paying attention to the internet and are seeking ways to curate at least parts of it in a coherent way.

19. Meg Leta Ambrose, "It's About Time: Privacy, Information Life Cycles, and the Right to Be Forgotten," *Stanford Technology Law Review* 16 (2013): 372. Data quoted from Daniel Gomes and Mario J. Silva, "Modelling Performance Persistence on the Web," *Proceedings of the 6th International Conference on Web Engineering* (2006). I have been unable to locate this source.

20. Ibid., 369–421; Jeffrey Rosen, "The Right to Be Forgotten," *Stanford Law Review Online* 16 (2012): 88–92 at http://www.stanfordlawreview.org/online/privacy-paradox/right-to-be-forgotten (accessed March 2, 2015). On May 13, 2014, the Court of Justice of the European Union found that Google is responsible for collecting information and disclosing that information to others, and so falls under the requirements of the European Union's data protection laws covering individual privacy: Google is supposed to seriously consider removing links to information about individuals that comes up during name searches that the individual finds "inadequate, irrelevant or no longer relevant, or excessive in relation to the purposes for which they were processed and in the light of the time that has elapsed," when that individual requests it. Court of Justice of the European Union, Press Release no. 70/14, Luxembourg, May 13, 2014, at curia.europa.eu/jcms/jcms/P_127116 (accessed March 2, 2015).

21. Ethicists are far from settled on how to apply privacy concerns to internet research. See Natasha Whiteman, *Undoing Ethics: Rethinking Practice in Online Research* (New York: Springer, 2012); Elizabeth A. Buchanan, *Readings in Virtual Research Ethics: Issues and Controversies* (Hershey, PA: Information Science Publications, 2004).

22. Secretary's Advisory Committee on Human Research Protections, "Considerations and Recommendations Concerning Internet Research and Human Subjects Research," posted as a pdf at www.hss-gov/ohrp/sachrp/commsec/ (accessed April 3, 2014); significant analysis of internet research and human subjects regulations appeared in Mark S. Frankel and Sanyin Siang, "Ethical and Legal Aspects of Human Subjects Research on the Internet," a report to the American Association for the Advancement of Science (1999), posted as a pdf at www.aaas.org/page/shrl-ethics-law-activities (accessed April 12, 2014). Since the late 1990s, there has been emerging literature on internet research and ethics, including the launch of the journal *Ethics and Information Technology.* For an important critique of the naïve application of the regulations on research using human subjects in internet research, see Joseph B. Walther,

"Research Ethics in Internet-enabled Research: Human Subjects Issues and Method-ological Myopia," *Ethics and Information Technology* 4 (2002): 205–212.

23. Discussion of the enormous challenges involved with preserving electronic records is beyond the scope of this project. It is daunting.

24. "SAA Core Values Statement and Code of Ethics" at www2.archivists.org/statements/saa-core-values-statement-and-code-of-ethics. The statement of values was approved by the SAA Council in 2011; the code of ethics was approved in 2005, and revised in 2012.

25. Marybeth Gaudette, "Playing Fair with the Right to Privacy," *Archival Issues* 28 (2003–2004): 23–24; Joan Hoff-Wilson, "Access to Restricted Collections: The Responsibility of Professional Historians," *The American Archivist* 46 (1983): 441–448.

26. "SAA Core Values Statement and Code of Ethics."

27. "SAA Core Values Statement and Code of Ethics." For a discussion of the challenges of dealing with culturally sensitive materials, see Alexa Roberts, "Trust Me, I Work for the Government: Confidentiality and Public Access to Sensitive Information," *American Indian Quarterly* 25 (2001): 13–17.

28. Verne Harris, "The Archival Sliver: Power, Memory and Archives in South Africa," *Archival Science* 2 (2002): 72n24. See also his *Archives and Justice: A South African Perspective* (Chicago, IL: Society of American Archivists, 2007) and his "Ethics and the Archive: 'An Incessant Movement of Recontextualization,'" in *Controlling the Past*, ed. Cook, 345–362.

29. Mark A. Greene, "A Critique of Social Justice as an Archival Imperative: What *Is* It We're Doing That's All That Important?" *The American Archivist* 76 (2013): 302–334; Randall C. Jimerson, "Archivists and Social Responsibility: A Response to Mark Greene," *The American Archivist* 76 (2013): 335–345. Some social activist archivists further argue that, in extreme conditions, archivists should sabotage records if doing so could prevent political torture and resist oppression; as Greene points out, this is a highly contested suggestion.

30. Danielson, *The Ethical Archivist*, 165–179.

31. Obviously, awareness of activist archivists comes from archivists willing to speak out about what they have done. See, for example, Shelley L. Davis, *Unbridled Power: Inside the Secret Culture of the IRS* (New York: HarperBusiness, 1997). Davis lost her job at the IRS after complaining about the destruction of records; no formal investigation was ever launched, however. In a more politically open case, archivists joined forces with legislators to argue fiercely that the records of the pro-segregationist Mississippi Sovereignty Commission be preserved after it was dissolved in 1972, and they were transferred to the state archives. Sarah Rowe-Sims, Sandra Boyd, and H. T. Holmes, "Balancing Privacy and Access: Opening the Mississippi State Sovereignty Commission Records," in *Privacy and Confidentiality Perspectives*, eds. Behrnd-Klodt and Wosh, 159–174.

32. As many commentators in the social justice literature admit, record keeping by the Nazis, the apartheid regime in South Africa, the Khmer Rouge, and other oppressive political states has been crucial for recognizing the extent of wrongdoing and, in some cases, prosecuting criminal behavior and returning stolen property. Greene, "A Critique of Social Justice as an Archival Imperative," 302–334.

33. Elena Danielson (Hoover Institution, Stanford University, retired), telephone interview, May 16, 2014. She acknowledged turning down such a collection during her time as archivist at the Hoover Institution, Stanford University.

34. Cook, ed., *Controlling the Past*; see Sue McKemmish, "Traces: Document, Record, Archive, Archives," in *Archives: Recordkeeping in Society*, eds. Sue McKemmish, Michael Piggott, Barbara Reed, and Frank Upward (Wagga Wagga, New South Wales: Center for Information Studies, 2005), 1–20, for an analysis of an event in Australia that illuminates various ways that an episode may be reconstructed from surviving traces.

35. Richard J. Cox, "The Documentation Strategy and Archival Appraisal Principles: A Different Perspective," in *American Archival Studies: Readings in Theory and Practice*, ed. Randall C. Jimerson (Chicago, IL: Society of American Archivists, 2000), 218.

36. Francis X. Blouin, Jr. "The Evolution of the Archival Practice and the History–Archival Divide," in *Controlling the Past*, ed. Cook, 317–328, summarizes the argument given in more detail in Blouin and Rosenberg, *Processing the Past*. Central to the argument that archivists must plan on what to save is the increasing separation of archivists from historians, and so the archival perspective from the historical one. See also, Boles and Greene, "Et Tu Schellenberg?" 298–310.

37. Mark Greene, "MPLP: It's Not Just for Processing Anymore," *The American Archivist* 73 (2010): 175–203.

38. "General Records Schedule 1: Civilian Personnel Records" on the National Archives and Records Administration General Records Schedules page at www.archives.gov/records-mgnt/grs/grs01.html (accessed February 10, 2014). Certain agencies are excluded. NARA is also required to review any records created before January 1, 1921, before they are destroyed, no matter their content.

39. Danielson, *The Ethical Archivist*, 87–117.

40. Paul Ohm, "Broken Promises of Privacy: Responding to the Surprising Failure of Anonymization," *UCLA Law Review* 57 (2009–2010): 1701–1777, provides a readable survey of recent work on re-identification of de-identified and "anonymized" datasets. For a case study of how archivists dealt with raw data gathered before IRB data destruction requirements, see Diane E. Kaplan, "The Stanley Milgram Papers: A Case Study on Appraisal of and Access to Confidential Data Files," *The American Archivist* 59 (1996): 288–297.

41. Malcolm Todd, "Power, Identity, Integrity, Authenticity, and the Archives: A Comparative Study of the Application of Archival Methodology to Contemporary Privacy," *Archivaria* 61 (2006): 181–215; quotation from 204.

42. Frank Boles, "Presidential Address: But a Thin Veil of Paper," *The American Archivist* 73 (2010): 20.

43. Three essays reprinted in *American Archival Studies: Readings in Theory and Practice*, ed. R. Jimerson, provide a good introduction to the range of approaches for planning acquisitions: Timothy L. Ericson, "At the 'Rim of Creative Dissatisfaction': Archivists and Acquisition Development," 177–182; Helen Willa Samuels, "Who Controls the Past?" 193–210; and Richard J. Cox, "The Documentation Strategy and Archival Appraisal Principles: A Different Perspective," 211–244.

44. "Saving Michigan's History—Preserving Personal Papers, Family Papers, and Records of Organizations: A Guide for New Donors," on the Bentley Historical Library website at bentley.umich.edu/mhchome/donors.php (accessed February 19, 2014.)

45. Ericson, "At the 'Rim of Creative Dissatisfaction,'" 181. Institutions with tightly defined purposes, such as the Leo Baeck Institute, Center for Jewish History, in New

York City, which collects "everything" to do with Jews and the Holocaust, justifiably cast a wide net. Michael Simonson (archivist, Leo Baeck Institute) telephone interview, May 27, 2014.

46. "American Heritage Center Acquisition Guidelines," n.d. and "AHC Manuscripts Collecting Policy, 15 October 2008," Policies and Academic Plan, American Heritage Center, University of Wyoming, at www.uwyo.edu/ahc/about/policies.html. Elena Danielson included the texts of these documents in Appendix B of her *The Ethical Archivist*, 340–371, as examples of well-thought, ethically-minded plans.

47. "American Heritage Center Acquisition Guidelines," n.d., pp. 2–3; Mark Greene, Director of the American Heritage Center, noted that financial records often involve third parties whose privacy could be compromised. Medical records, in addition to raising privacy concerns, can also be very uninformative and so not worth collecting. An exception would likely be made for the medical records of a significant historical actor known to have had health concerns. Mark Greene, telephone interview, April 27, 2014.

48. Interview with an archivist at a research university; this informant asked to remain anonymous. Archivists have not yet researched the question of how often privacy concerns prevent the acquisition of records, especially the acquisition of patient records.

49. For a discussion of post-mortem donor restrictions, see Danielson, *The Ethical Archivist*, 185–208; Raymon H. Geselbracht, "The Origins of Restrictions on Access to Personal Papers at the Library of Congress," *The American Archivist* 49 (1986): 142–162; for an important discussion of the processing of controversial government files, see Rowe-Sims, Boyd, and Holmes, "Balancing Privacy and Access," 159–174.

50. Mark A. Greene and Dennis Meissner, "More Product, Less Process: Revamping Traditional Archival Processing," *The American Archivist* 68 (2005): 210. The processing of digital collections is also time-consuming if each file has to be checked.

51. Ibid., 252.

52. Bruce P. Stark, *A Guide for Processing Manuscript Collections*, quoted in Greene and Meissner, "More Product, Less Process," 252n124.

53. Simonson, telephone interview, May 27, 2014.

54. Danielson, *The Ethical Archivist*, 211.

55. Interview with an archivist at a major Eastern research institution, summer 2014. This archivist asked to remain anonymous.

56. The law is very clear that donors own the items sent to them from others, and all items that they themselves created. Donors do not own the copyright to the content of the items that other people have written, however, and archives and libraries are very clear that users must get permission from the author of items in order to quote from them. Danielson, *The Ethical Archivist*, 141; Menzi L. Behrnd-Klodt, *Navigating Legal Issues in the Archives* (Chicago, IL: Society of American Archivists, 2008), 203–248, has a full discussion of copyright law for archivists.

57. Gaudette, "Playing Fair with the Right to Privacy," 21.

58. Mark A. Greene, "Moderation in Everything, Access in Nothing? Opinions about Access Restrictions on Private Papers," *Archival Issues* 18 (1993): 31–41; one of the most vocal proponents of second- and third-party privacy is MacNeil, *Without Consent*.

59. Gaudette, "Playing Fair with the Right to Privacy," 28, 31.

60. MacNeil, *Without Consent*, 116–118; see also Jill Cariffe Cirasella, "At Odds?: Archives and Privacy," *Current Studies in Librarianship* 24 (2000): 88–92.

61. Gaudette, "Playing Fair with the Right to Privacy," 31.

62. For the possibility that all research information about living individuals could fall under regulations for the protections of human subjects, see chap. 6. Some state laws do have specific time limits on closure built into their public records laws. See Bauman, "The Administration of Access to Confidential Records in State Archives," 365, for a discussion of Georgia's "model" law that closes confidential case files for seventy-five years after the date of creation.

63. Kaplan, "The Stanley Milgram Papers," 290–292; Mrs. Milgram left it up to the Yale archives staff to decide which parts of the collection were confidential files and which were not, with the former to be closed for seventy-five years; the archives staff, after much consultation, decided that files about all of Milgram's experiments in which subjects were identified would be closed for seventy-five years after the date of the experiment.

64. V.J.H. Cain, "The Ethics of Processing," *Provenance* 11 (1993): 39–55.

65. Judith Schwarz, "The Archivist's Balancing Act: Helping Researchers While Protecting Individual Privacy," *The Journal of American History* 79 (1992): 183–184; Amy Fitch (The Rockefeller Archives), telephone interview, July 1, 2014.

66. For important comments on the vital role that collection description plays in providing access to materials, see Boles, "Presidential Address: But a Thin Veil of Paper," 22–23; some archivists note with disapproval that omitting the existence of restricted files from finding aids continues to be an acceptable practice in some archives; Greene, telephone interview, April 27, 2014; Simonson, telephone interview, May 27, 2014.

67. Julie Herrada, "Letters to the Unabomber: A Case Study and Some Reflections," *Archival Issues* 28 (2003–2004): 41.

68. Entry on the Ted Kaczynski papers in the University of Michigan's online catalog at http://mirlyn.lib.umich.edu/Record/004130546/Description#summary (accessed February 27, 2014). The Mississippi state government paid for a more complex indexing and redaction process for the records of the Mississippi Sovereignty Commission papers, using electronic scanned copies. It should be noted that it took twenty years of litigation before a compromise was worked out on how the records could be used while protecting the privacy of named victims. Rowe-Sims, Boyd, and Holmes, "Balancing Privacy and Access," 159–174.

69. See the finding aid for the Chellis Glendinning Papers (1890–2008, bulk 1980s–2007), via the link at http://mirlyn.lib.umich.edu/Record/005939917. There are five boxes of correspondence, much of it from the period from the late 1970s to 2008, all of which are listed as completely open. The quotation comes from the Biography page of the finding aid.

70. Greene, "Moderation in Everything, Access in Nothing?" 41; Novak, telephone interview, May 21, 2014.

71. Danielson, *The Ethical Archivist*, 213.

72. Danielson, telephone interview, May 16, 2014.

73. See the cases contributed by Mark Greene and Robert Sink in Karen Benedict, *Ethics and the Archival Profession: Introduction and Case Studies* (Chicago, IL: Society of the American Archivists, 2003), 64–66; note that Benedict wrote the commentary on the cases.

74. Behrnd-Klodt, *Navigating Legal Issues in the Archives*, 104, 107, 111–112; Timothy D. Pyatt, "Southern Family Honor Tarnished? Issues of Privacy in the Walker Percy and Shelby Foote Papers," in *Privacy and Confidentiality Perspectives*, eds. Behrnd-Klodt and Wosh, 149–158.

75. Danielson, *The Ethical Archivist*, 125–137.

76. Steven Bingo, "Of Provenance and Privacy: Using Contextual Integrity to Define Third-Party Privacy," *The American Archivist* 74 (1993): 31–41.

77. Arminda Bradford Bepko, "Public Availability or Practical Obscurity: The Debate over Public Access to Court Records on the Internet," *New York Law School Law Review* 49 (2004/2005): 967–991; "Public Records on the Internet: The Privacy Dilemma" (2002), posted on the Privacy Rights Clearinghouse: Empowering Consumers, Protecting Privacy website at https://www.privacyrights.org/ar/onlinepubrecs .htm (accessed March 19, 2014). Complaints that Google Street View photographs violate privacy have not been upheld in courts when the photographs are taken from public property and are of publicly visible structures. See also Dean Seeman, "Naming Names: The Ethics of Identification in Digital Library Metadata," *Knowledge Organization* 39 (2012): 325–331; Mark Tunick, "Privacy and Punishment," *Social Theory and Practice* 39 (2013): 656.

78. Bepko, "Public Availability or Practical Obscurity," 967–991.

79. Dejah T. Rubel, "Accessing their Voices from Anywhere: Analysis of the Legal Issues Surrounding the Online Use of Oral Histories," *Archival Issues* 31 (2007): 171–187, especially 181; Casey S. Westerman, "Last Words: Suicide Notes, Ownership, Access and Privacy," paper presented at the annual meeting of the Society of American Archivists, Washington, DC, August 15, 2014.

80. Bauman, "The Administration of Access to Confidential Records in State Archives," 349–369.

81. Danielson, *The Ethical Archivist*, 142. For an important case involving a donor who screened research use, see Harold L. Miller, "Will Access Restrictions Hold Up in Court? The FBI's Attempt to Use the Braden Papers at the State Historical Society of Wisconsin," *The American Archivist* 52 (1983): 180–190.

82. Greene, telephone interview, April 27, 2014.

83. See The Huntington "Admission to the Research Library" at http://www.huntington .org/WebAssets/Templates/content.aspx?id=586 (accessed March 5, 2014).

84. The Rockefeller Archive, for example, does not allow only certain researchers to access collections. As Amy Fitch, an archivist there put it, "closed is closed." Fitch, telephone interview, July 1, 2014.

85. Minnesota History Center, "Application to Restricted Records in State Archives," a pdf file posted at "FAQ—Restricted Collections," at http://sites.mnhs.org/library/ content/faq-restricted-collections (accessed March 13, 2014).

86. Wiener, interview, Columbus, OH, June 16, 2014.

87. Georgia Archives, "Georgia Records Act" posted as a pdf at www.georgiaarchives .org/records/laws (accessed March 13, 2014). I have not found any legal cases that reached an appellate level where archives or users have been sued for privacy, or where users have had actions taken against them under state law for violating user agreements, and so have no sense of how the courts have dealt with researcher access to confidential information held in archives. When asked, none of the archivists interviewed for this project could come up with any examples of archives

being sued, although a few faced threats of lawsuits. Danielson, telephone interview, May 16, 2014; Fitch, telephone interview, July 1, 2014.

88. Danielson, *The Ethical Archivist*, 141; Behrnd-Klodt, *Navigating Legal Issues in the Archives*, 204–264, especially 238.

89. Herrada, "Letters to the Unabomber," 42; theoretically, an archive might be able to sue a researcher for breach of contract if the researcher violates a user agreement and the court agrees that the agreement is a contract. The literature that I have examined on archives and legal issues makes no mention of this possibility, and no archivist I have interviewed knew of a case where an archive took this step.

90. Danielson, *The Ethical Archivist*, 211–212; Mark Greene and Christine Weideman, "The Buckley Stops Where? The Ambiguity and Archival Implications of the Family Educational Rights and Privacy Act," in *Privacy and Confidentiality Perspectives*, eds. Behrnd-Klodt and Wosh, 181–198, suggest that IRB review would be a good approach to researching FERPA covered materials. Kaplan, "The Stanley Milgram Papers," 293, rejected the use of a review board for screening access to Milgram's papers because she thought that the library would have to establish a review committee and would not have the necessary expertise to evaluate proposals from different academic disciplines. She did not discuss trying to use Yale's social and behavioral sciences IRB for this purpose.

91. Susan C. Lawrence, "Access Anxiety: HIPAA and Historical Research," *Journal of the History of Medicine and Allied Sciences* 62 (2007): 422–460; all of the information in this article is still valid, although it now only applies to the records of the deceased for fifty years.

92. The privacy officer of a covered institution may also review and approve research that only involves the records of the deceased. Ibid., 440–446.

93. Ibid., 436–460.

94. Privacy Board, The Johns Hopkins Medical Institutions, "Application for a Waiver of Authorization for Research Use or Disclosure of Protected Health Information (PHI) and Other Personal Information that is Protected by Law," pdf file at http://www.medicalarchives.jhmi.edu/hipaaform.html (accessed March 17, 2014).

95. Augustus C. Long Health Sciences Library, Columbia University, "Access Policies: Access to Records Containing Protected Health Information," at http://vesta.cumc.columbia.edu/library/archives/accesspatient.html (accessed March 18, 2014); Novak, telephone interview, May 21, 2014.

96. Lawrence, "Access Anxiety," 435; Major Charles G. Kels, "Privacy after Death?" *Reporter* 38 (2011): 36–40, explains how the Department of Defense has adopted the Privacy Rule for military medical records. The National Library of Medicine also created a HIPAA-like policy, in that they require an application to use patient records and assurances that no names will be published. The NLM policy applies to records for only one hundred years after the date of creation, however. John Rees (NLM archivist), personal communication, Bethesda, MD, June 28, 2011.

97. Phoebe Evans Letocha and Emily R. Novak Gustainis, "Recommended Practices for Enabling Access to Manuscript and Archival Collections Containing Health Information about Individuals," available on the Medical Heritage Library website at http://www.medicalheritage.org/2015/02/now-available-recommended-practices-for-enabling-access-to-manuscript-and-archival-collections-containing-health-information-about-individuals/.

Chapter 5 — Managing Privacy: Historians at Work

1. Interviews with two archivists who wished to remain anonymous; neither would give any details about the events that concerned them. In neither case did they know if the individuals discussed by researchers were actually offended or upset. The archivists were unhappy on behalf of those whose materials were deposited in their collections.

2. Jon Wiener, *Historians in Trouble: Plagiarism, Fraud, and Politics in the Ivory Tower* (New York: The New Press, 2005); Ron Robin, *Scandals & Scoundrels: Seven Cases that Shook the Academy* (Berkeley: University of California Press, 2004); see also Thomas Mallon, *Stolen Words: The Classic Book on Plagiarism*, 2nd ed. (San Diego: Harcourt, Inc., 2001). Historians certainly plagiarize and commit fraud, so why would they not breach confidences and invade privacy?

3. Consider two biographies of Thomas Jefferson: Thomas E. Watson, *Life and Times of Thomas Jefferson* (New York: D. Appleton and Co., 1903), which does not mention Sally Hemmings at all, and Andrew M. Allison, *The Real Thomas Jefferson. Part I: Thomas Jefferson: Champion of Liberty* (Washington, DC: National Center for Constitutional Studies, 1983), which denied all scurrilous rumors about any improper relations between Hemmings and Jefferson.

4. Diane Wood Middlebrook, *Anne Sexton: A Biography* (Boston: Houghton Mifflin, 1991). Diane Wood Middlebrook, "Telling Secrets," in Mary Rhiel and David Suchoff, *The Seductions of Biography* (New York: Routledge, 1996), 124–129.

5. Linda Grey Sexton, quoted in Middlebrook, "Telling Secrets," 126.

6. Ibid., 127.

7. Middlebrook, *Anne Sexton*, 66.

8. Martha Stephens, *The Treatment: The Story of Those Who Died in the Cincinnati Radiation Tests* (Durham, NC: Duke University Press, 2002), 12.

9. Advisory Committee on Human Radiation Experiments, *The Human Radiation Experiments* (New York: Oxford University Press, 1996).

10. Stephens, *The Treatment*, 21–26.

11. Ibid., 293–295. All of the names are listed on the plaque and in Stephens's book, even those for whom no family came forward to be included in the lawsuit.

12. Ibid., 23.

13. Ibid., 26.

14. Janna Malamud Smith, *Private Matters: In Defense of the Personal Life* (Reading, MA: Addison-Wesley, 1997), 3–6, 145–172.

15. Constance Putnam (independent scholar), telephone interview, July 21, 2014.

16. See chap. 2.

17. Susan Reverby (Wellesley College), telephone interview, July 24, 2014.

18. Not all historians who talk to people as part of their research intend that their interviews will end up being shared with others as formal oral history, of course. When acting like anthropologists or sociologists, historians may dig more deeply into private information and possibly self-incriminating testimony in exchange for assurances that identities will be protected. Charles L. Bosk provides an important critique of the ethics of ethnography in his "Irony, Ethnography, and Informed Consent," in Barry Hoffmaster, ed., *Bioethics in Social Context* (Philadelphia, PA: Temple University Press, 2001), 199–220. His ethnographic subjects felt considerable betrayal over the way he reported on their ideas and actions in his work. They simply could not be anonymized sufficiently to be unrecognizable in the communities in which he carried out his research.

19. Johanna Schoen (Rutgers University), interview, Iowa City, IA, June 8, 2011; Johanna Schoen, *Choice & Coercion: Birth Control, Sterilization and Abortion in Public Health and Welfare* (Chapel Hill: The University of North Carolina Press, 2005).

20. Schoen, interview, June 8, 2011.

21. Victims of 1929–1974 N.C. Eugenics Program urged to contact N.C. DHHS, at www.ncdhhs.gov/pressrel/2008/2008–15–5-eugenics.htm (accessed August 21, 2014); see the Office of Justice for Sterilization Victims, supported by the Justice for Sterilization Victims Foundation, at www.sterilizationvictims.nc.gov (accessed August 21, 2014), which has posted copies of meeting minutes, reports, request forms, and news clippings.

22. Telephone interview with a historian who asked to remain unidentified.

23. I found a wonderful box of individuals' life insurance medical examination reports from about 1910 to the early 1950s in an archive. The finding aid only indicated that it held information about life insurance. It was not at all useful for my research at the time, so I closed it up and had it returned to storage. Should this material be restricted? I did not ask.

24. Interview with an archivist at a major research institution who asked to remain anonymous.

25. See chap. 3; presumably a highly significant revision to historical understanding would garner media attention. If deemed newsworthy, it could be very hard for a person to win a tort case against the historian.

26. For an example, see Randolph Roth, *American Homicide* (Cambridge, MA: Belknap Press of Harvard University Press, 2009); Eric C. Schneider (University of Pennsylvannia), interview, Bala Cynwyd, PA, June 16, 2011: "I am trained as a social historian, I'm interested in broad patterns, in how institutions work . . . so whether someone was named Mary Smith or Mary Jones makes absolutely no difference to me."

27. There is an extensive range of materials on narrative theory that discusses the ethics of identity and relationships in literature, almost entirely in discussions of fiction. See, for some of the most prominent examples, James Phelan, *Living to Tell About It: A Rhetoric and Ethics of Character Narration* (Ithaca, NY: Cornell University Press, 2005); James Phelan, *Narrative as Rhetoric: Technique, Audiences, Ethics, Ideology* (Columbus, OH: Ohio State University Press, 1996; James Phelan, ed., *Reading Narrative: Form, Ethics, Ideology* (Columbus, OH: Ohio State University Press, 1989); Karl Simms, ed., *Ethics and the Subject. Critical Studies*, 8 (Amsterdam: Rodopi, 1997); Meili Steele, *Theorizing Textual Subjects: Agency and Oppression. Literature, Culture, Theory*, vol. 21 (New York: Cambridge University Press, 1997). For one of the few examples of work about non-fiction, see Kay Schaffer and Sidonie Smith, eds., *Human Rights and Narrated Lives: The Ethics of Recognition* (New York: Palgrave Macmillan, 2004), which primarily deals with the appropriation of personal stories for political purposes. For a discussion of the more immediate problems with insufficient de-identification in ethnographies, particularly ethnographies done among physicians and counselors in hospital settings, see Bosk, "Irony, Ethnography, and Informed Consent," 199–220.

28. See the discussion of *Haynes v. Alfred A. Knopf, Inc. and Nicolas Lemann* (1993) in chap. 3, for a jurist's eloquent observations on this point.

29. For an important and much more sophisticated analysis of the epistemological work done by historical narratives than I could possibly offer, see Allan Megill, *Historical*

Knowledge, Historical Error: A Contemporary Guide to Practice (Chicago, IL: University of Chicago Press, 2007), chaps. 3 and 4.

30. David Gary Shaw, "Happy in Our Chains? Agency and Language in the Postmodern Age," *History and Theory* 40 (2001): 8. This theme issue, "Agency after Postmodernism," contains important articles on theories of agency among theories of discourse, language, and history. All provide perspectives that help to return individual action to history without simplistically rejecting the insights of postmodernism on social thought. For a recent discussion of agency in early modern history, see Cornelia Hughes Dayton, "Rethinking Agency, Recovering Voices," *American Historical Review* 109 (2004): 827–843; James C. Scott, *Domination and the Arts of Resistance: Hidden Transcripts* (New Haven, CT: Yale University Press, 1990) provides many examples of how to discern agency among the dispossessed; Ann Laura Stoler, *Along the Archival Grain: Epistemic Anxieties and Colonial Common Sense* (Princeton, NJ: Princeton University Press, 2009) similarly complicates simple dichotomies of power/subjugation, although with little interest in identifying individual subjects who resisted Dutch colonial projects; Warwick Anderson, *The Collectors of Lost Souls: Turning Kuru Scientists into Whitemen* (Baltimore, MD: The Johns Hopkins University Press, 2008) offers an important example from the mid-twentieth century.

31. It can be challenging to find exactly where authors acknowledge or explain the use of pseudonyms. See, for example, *Warwick Anderson, The Collectors of Lost Souls: Turning Kuru Scientists into Whitemen* (Baltimore, MD: Johns Hopkins University Press, 2008), 59–67, 248n1–249n24. If the names of the Fore people sick with kuru mentioned in the text are pseudonyms, this is not specified in the respective notes.

32. Elizabeth Hampsten, *Read This Only to Yourself: The Private Writings of Midwestern Women, 1880–1910* (Bloomington: Indiana University Press, 1982), ix, xi–xii; see especially chap. 4.

33. Peter Wallenstein, *Tell the Court I Love My Wife: Race, Marriage, and Law—An American History* (New York: Palgrave Macmillan, 2002).

34. Janet Golden, *Message in a Bottle: The Making of Fetal Alcohol Syndrome* (Cambridge, MA: Harvard University Press, 2005), chap. 7; Janet Golden (Rutgers University–Camden), interview, Bala Cynwyd, PA, June 16, 2011.

35. Jacki Rand (University of Iowa), telephone interview, June 24, 2014.

36. Leslie Reagan, *When Abortion was a Crime: Women, Medicine and Law in the United States, 1867–1973* (Berkeley: University of California Press, 1997), 255–256.

37. Leslie Reagan (University of Illinois, Champagne–Urbana), telephone interview, June 17, 2014.

38. Robert Lilly (Northern Kentucky University), telephone interview, August 8, 2014. U.S. publishers in fact refused to touch Lilly's manuscript. No one wanted to "besmirch the Greatest Generation." His book was first published in France, and then in the U.K. While Robert Lilly is a sociologist, his work is squarely situated within historical research. J. Robert Lilly, *Taken By Force: Rape and American GIs in Europe During World War II* (Houndsmills, Hampshire, UK: Palgrave Macmillan, 2007).

39. Reverby, interview, July 24, 2014; Reverby provided citations to the case records where the real names can be found. Susan Reverby, *Examining Tuskegee: The Infamous Syphilis Study and its Legacy* (Chapel Hill: University of North Carolina Press, 2009).

40. Reagan, interview, June 17, 2014.

41. Reagan, interview, June 17, 2014; Reverby, interview, July 24, 2014; Lilly, interview, August 8, 2014.

42. Reagan, interview, June 17, 2014.

43. Jack Pressman, *Last Resort: Psychosurgery and the Limits of Medicine* (San Francisco, CA: University of California Press, 1998), 498n2.

44. Christopher Crenner (University of Kansas), telephone interview, June 18, 2014. In his article about health and race in Kansas in the early twentieth century, Crenner of course did not use real names of patients. He did, however, provide detailed citations to his archival material so that names could be recovered by other researchers if necessary. Christopher Crenner, "Race and Medical Practice in Kansas City's Free Dispensary," *Bulletin of the History of Medicine* 82 (2008): 821n2.

45. Emily Abel, *Hearts of Wisdom: American Women Caring for Kin, 1850–1940* (Cambridge, MA: Harvard University Press, 2000), 150–154; Emily Abel (University of California at Los Angeles), telephone interview, July 18, 2014; for the restrictions on the collections of the Charity Organization Society, see "Access Restrictions," in the finding aid for the Community Service Society Archives, 1842–1995, Columbia University Library, at http://findingaids.cul.columbia.edu/ead//nnc-rb/ldpd_4079675#using_collection (accessed November 19, 2014).

46. Analysis of this point could be greatly expanded by making the obvious parallel with reading texts about other subordinated peoples: African Americans, Native Americans, colonial subjects, other non-Western, marginalized people. Doing so goes well beyond the scope of this book; for initial inspiration, see bell hooks, "Marginality as a Site of Resistance," in *Out There: Marginalization and Contemporary Cultures*, eds., Russell Ferguson, et al. (Cambridge, MA: MIT Press, 1990), 241–243.

47. There are, of course, exceptions. Jack Pressman used pseudonyms for the names of doctors associated with the case files he used in *Last Resort*, 498n2, although he did not say if this was to protect the doctors or to further obscure possible identification of the patients.

48. Gerald E. Markowitz and David Rosner, *Deceit and Denial: The Deadly Politics of Industrial Pollution* (Berkeley: University of California Press, 2002). See, for example, the details about Verald K. Rowe, a biochemist who worked for Dow Chemical, 172, 351n18.

49. Sydney Halpern (University of Illinois at Chicago), interview, Chicago, IL, January 8, 2012; revised by email March 17, 2015.

50. I have deliberately omitted the current status of Halpern's decision from this discussion.

51. See chap. 2.

52. Alice Wexler, *The Woman Who Walked into the Sea: Huntington's and the Making of a Genetic Disease* (New Haven, CT: Yale University Press, 2008), xxi–xxii.

53. Megill, *Historical Knowledge, Historical Error*, 17–59.

54. Victor Jeleniewski Seidler, "Obligations to the Dead: Historical Justice and Cultural Memory," *European Judaism* 46 (2013): 12–31, and Anita Shapira, "The Holocaust: Private Memories, Public Memory," *Jewish Social Studies* n.s. 4 (1998): 40–58 provide examples of history/memory concerns in Holocaust literature, where much has been written. In American history, the history/memory collisions appear largely in political discussions about what kind of histories the public should support and children should be taught. See, for an important example, Edward T. Linenthal and

Tom Engelhardt, *History Wars: The Enola Gay and Other Battles for the American Past* (New York: Henry Holt, 1996).

55. Richard J. Bernstein, "The Culture of Memory; review essay on *The Ethics of Memory* by Avishai Margalit," *History and Theory* 43 (2004): 168, emphasis original. For one of the many articles on the political meaning of memory and memorialization, see Kevin Bruyneel, "The King's Body: The Martin Luther King Jr. Memorial and the Politics of Collective Memory," *History and Memory* 26 (2014): 75–108.

56. Anna Sheftel and Stacey Zembrzycki, "Only Human: A Reflection on the Ethical and Methodological Challenges of Working with 'Difficult' Stories," *The Oral History Review* 37 (2010): 191–214; Bernstein, "The Culture of Memory," 166–167.

57. Michael Bergman, "General Was Ohioan Despite Side in War," *Columbus Dispatch* (Ohio), January 7, 2015 (Letter to the Editor), online at http://www.dispatch.com/content/stories/editorials/2015/01/07/1-general-was-ohioan-despite-side-in-war.html (accessed February 1, 2015); Didier Fassin, "The Humanitarian Politics of Testimony: Subjectification through Trauma in the Israeli–Palestinian Conflict," *Cultural Anthropology* 23 (2008): 531–558; Schaffer and Smith, *Human Rights and Narrated Lives*; Roger I. Simon, *The Touch of the Past: Remembrance, Learning, Ethics* (New York: Palgrave Macmillan, 2005); Gina Marie Weaver, *Ideologies of Forgetting: Rape in the Vietnam War* (Albany: State University of New York Press, 2010).

58. Avishai Margalit, *The Ethics of Memory* (Cambridge, MA: Harvard University Press, 2002), 18–26.

59. Ibid., 24–26, 91–95.

60. Ibid., 20, 93.

61. Sigmund Freud, *Reflections on War and Death*, translated from the German 1915 essay by A. A. Brill and Alfred B. Kuttner (New York: Moffat, Yard, and Co., 1918), 16.

62. E. Partridge, "Posthumous Ethics and Posthumous Respect," *Ethics* 91 (1981): 261.

63. Daniel Sperling, *Posthumous Interests: Legal and Ethical Perspectives* (New York: Cambridge University Press, 2008), 239; Kirsten Rabe Smolensky, "Rights of the Dead," *Hofstra Law Review* 37 (2009): 763–803.

64. Sperling, *Posthumous Interests*, 8–47; quotation 9. Sperling tries to use interests to establish medical confidentiality for the dead during his analysis of the inadequacies of rights-based arguments and his survey of medical confidentiality provisions in various laws. His analysis is quite tentative, however, and in the end he retreats from any absolute statement about the long-term confidentiality of medical records. Indeed, he suggests that disclosure might be appropriate "where this would serve to correct a serious error and rehabilitate the person's reputation" (233). See Sperling's references for a way into the technical philosophical literature about the rights and interests of the dead.

65. Malin Masterton, Gert Helgesson, Anna T. Höglund, and Mats G. Hansson, "Queen Christiana's Moral Claim on the Living: Justification of a Tenacious Moral Intuition," *Medicine, Health Care and Philosophy* 10 (2007): 321–327.

66. Søren Holm, "The Privacy of Tutankhamen—Utilizing the Genetic Information in Stored Tissue Samples," *Theoretical Medicine* 22 (2001): 447.

67. Ibid.; Zahi Hawass, et al. "Ancestry and Pathology in King Tutankhamun's Family," *JAMA* 303 (February 17, 2010), 638–647. Tutankhamen was the son of Akhenaten and Akhenaten's sister, but Nefertiti was not Akhenaten's sister, which was already known through other sources.

68. Masterton, Helgesson, Höglund, and Hansson, "Queen Christiana's Moral Claim on the Living," 321–327.

69. For work that argues that ethical decision-making should be far less based on abstract principles and far more on the judgment of those who understand the context (be that an individual or a team), see Renee R. Anspach and Diane Beeson, "Emotions in Medical and Moral Life," in Hoffmaster, ed., *Bioethics in Social Context*, 112–136; Renee C. Fox and Judith P. Swazey, "Medical Morality is not Bioethics—Medical Ethics in China and the United States," *Perspectives in Biology and Medicine* 27 (1984): 336–360; Barry Hoffmaster, "Can Ethnography Save the Life of Medical Ethics?" *Social Science & Medicine* 35 (1992): 1421–1431; Arthur Kleinman, "Anthropology of Bioethics," in Arthur Kleinman, ed., *Writing at the Margin: Discourse Between Anthropology and Medicine* (Berkeley: University of California Press, 1995), 41–67; Jose Lopez, "How Sociology Can Save Bioethics . . . Maybe," *Sociology of Health & Illness* 26 (2004): 875–896; Arthur Kleinman, Renée C. Fox, and Allan M. Brandt, eds., *Bioethics and Beyond, Daedalus* 128 (1999): 1–326, especially the articles by Charles Rosenberg, Veena Das, Charles L. Bosk, and Daniel Callahan; Natasha Whiteman, *Undoing Ethics: Rethinking Practice in Online Research* (New York, Dordrecht, Heidelberg, London: Springer, 2012), 47–80.

Chapter 6 — Conclusion: Resistance

1. Eric T. Dean, Jr., "Reflections on 'The Trauma of War' and *Shook Over Hell*," *Civil War History* 59 (2013): 414–418; Matthew Warshauer and Michael Sturges, "Difficult Hunting: Accessing Connecticut Patient Records to Learn About Post-Traumatic Stress Disorder during the Civil War," *Civil War History* 59 (2013): 419–452.

2. The text of "Human Subjects Research Protections: Enhancing Protections for Research Subjects and Reducing Burden, Delay and Ambiguity for Investigators" may be found at http://www.regulations.gov/#!documentDetail;D=HHS-OPHS-2011 -0005-0001. I submitted comments on the proposed rule on October 25, 2011. These may be viewed at http://www.regulations.gov/#!documentDetail;D=HHS-OPHS -2011-0005-0630.

3. For a more general critique of the IRB process and thoughts inspired by the 2011 ANPRM, see American Association of University Professors, "Regulation of Research on Human Subjects: Academic Freedom and the Institutional Review Board," (2013) at http://www.aaup.org/report/regulation-research-human -subjects-academic-freedom-and-institutional-review-board.

4. Antoon De Baets, "A Declaration of the Responsibilities of Present Generations toward Past Generations," *History and Theory* 43 (2004): 134, 136–137.

5. Ibid., 148, 158, 158n95.

6. Ibid., 143–144.

7. Ibid., 140–141.

8. Steven Kern, *The Culture of Love: Victorians to Moderns* (Cambridge, MA: Harvard University Press, 1992), 395.

9. See chap. 2, note 43.

10. François-Marie d'Arouet, "Lettres à M. De Genonville, Lettre Première (1710)," *Œuvres Complete de Voltaire*, vol. 1 (Paris: De L'Imprimerie de la Société Littéraire-Typographique, 1785), 28.

Chapter 6 — Conclusion: Resistance

Bibliography

Abbott v. Texas Dep't of Mental Health, 212 S.W. 648 (Tex. App.–Austin 2006, no pet.).

Abel, Emily. *Hearts of Wisdom: American Women Caring for Kin, 1850–1940*. Cambridge, MA: Harvard University Press, 2000.

Acara v. Banks, 470 F.3d 569, 571 (5th Cir. 2006).

"Access Restrictions." Charity Organization Society, Community Service Society Archives, 1842–1995. Columbia University Library. http://findingaids.cul.columbia.edu/ead//nnc-rb/ldpd_4079675#using_collection.

Advisory Committee on Human Radiation Experiments. *The Human Radiation Experiments*. New York: Oxford University Press, 1996.

Allison, Andrew M. *The Real Thomas Jefferson. Part I: Thomas Jefferson: Champion of Liberty*. Washington, DC: National Center for Constitutional Studies, 1983.

Alvarado, et al. v. KOB-TV, 493 F.3d 1210 (10th Cir. 2007).

Ambrose, Meg Leta. "It's About Time: Privacy, Information Life Cycles, and the Right to Be Forgotten." *Stanford Technology Law Review* 16 (2013): 369–421.

Amdur, Robert J. *Institutional Review Board Member Handbook*. Sudbury, MA: Jones and Bartlett, 2003.

American Association of University Professors. "Regulation of Research on Human Subjects: Academic Freedom and the Institutional Review Board." 2013. http://www.aaup.org/report/regulation-research-human-subjects-academic-freedom-and-institutional-review-board.

———. "Statement of Principles on Academic Freedom and Tenure." 1940. http://www.aaup.org/AAUP/pubsres/policydocs/contents/1940statement.htm.

American Heritage Center, University of Wyoming. "AHC Manuscripts Collecting Policy." October 15, 2008. http://www.uwyo.edu/ahc/about/policies.html.

———. "American Heritage Center Acquisition Guidelines." n.d. http://www.uwyo.edu/ahc/about/policies.html.

Anderson, Warwick. *The Collectors of Lost Souls: Turning Kuru Scientists into Whitemen*. Baltimore, MD: The Johns Hopkins University Press, 2008.

Awad, Isabel. "Journalists and Their Sources: Lessons from Anthropology." *Journalism Studies* 7, no. 6 (2006): 922–939.

Barritt, Marjorie Rabe. "The Appraisal of Personally Identifiable Student Records." *The American Archivist* 49 (1986): 263–275.

Bauman, Roland M. "The Administration of Access to Confidential Records in State Archives: Common Practices and the Need for a Model Law." *The American Archivist* 49 (1986): 349–369.

Behrnd-Klodt, Menzi L. *Navigating Legal Issues in the Archives* Chicago, IL: Society of American Archivists, 2008.

Behrnd-Klodt, Menzi L., and Peter J. Wosh, eds. *Privacy and Confidentiality Perspectives: Archivists and Archival Records*. Chicago, IL: Society of American Archivists, 2005.

Benedict, Karen. *Ethics and the Archival Profession: Introduction and Case Studies*. Chicago, IL: Society of The American Archivists, 2003.

Bentley Historical Library. "Saving Michigan's History—Preserving Personal Pa-

pers, Family Papers, and Records of Organizations: A Guide for New Donors."
http://bentley.umich.edu/mhchome/donors.php.

Bepko, Arminda Bradford. "Public Availability or Practical Obscurity: The Debate over
Public Access to Court Records on the Internet." *New York Law School Law Review* 49
(2004/2005): 967–991.

Berg, Jessica Wilen. "Grave Secrets: Legal and Ethical Analysis of Postmortem Confidentiality." *Connecticut Law Review* 34 (2001): 81–122.

Bergman, Michael. "General Was Ohioan Despite Side in War." *Columbus Dispatch*
(Ohio), January 7, 2015 (Letter to the Editor). http://www.dispatch.com/content/
stories/editorials/2015/01/07/1-general-was-ohioan-despite-side-in-war.html.

Bernstein, Richard J. "The Culture of Memory: Review essay on *The Ethics of Memory* by
Avishai Margalit." *History and Theory* 43 (2004): 165–178.

Binder, William H. "Publicity Rights and Defamation of the Deceased: Resurrection or
R.I.P." *DePaul-LCA Journal of Art and Entertainment Law* 12 (2002): 297–316.

Bingo, Steven. "Of Provenance and Privacy: Using Contextual Integrity to Define Third-
Party Privacy." *The American Archivist* 74 (1993): 31–41.

Blouin, Francis X., Jr., and William G. Rosenberg. *Processing the Past: Contesting Authority in History and the Archives*. New York: Oxford University Press, 2011.

"Board of Governors of the Federal Reserve System." Statistics and Historical Data.
Accessed August 31, 2011. http://www.federalreserve.gov/econresdata/releases/
statisticsdata.htm.

Boles, Frank. "Presidential Address: But a Thin Veil of Paper." *The American Archivist*
73 (2010): 19–25.

Boles, Frank, and Mark Greene. "Et Tu Schellenberg? Thoughts on the Dagger of American Appraisal Theory." *The American Archivist* 59 (1996): 298–310.

Bollinger, Lee C. *Uninhibited, Robust, and Wide-Open: A Free Press for a New Century.*
New York: Oxford University Press, 2010.

Bosk, Charles L., and Raymond G. DeVries. "Bureaucracies of Mass Deception: Institutional Review Boards and the Ethics of Ethnographic Research." *Annals of the American Academy* 595 (2004): 249–263.

"Boston College Subpoena News." Accessed January 31, 2014. http://
bostoncollegesubpoena.wordpress.com.

Bradley, Matt. "Silenced For Their Own Protection: How the IRB Marginalizes Those
It Feigns to Protect." *ACME: An International E-Journal for Critical Geographies* 6
(2007): 339–349.

Bradshaw, John. *Family Secrets: What You Don't Know Can Hurt You.* London: Piatkus,
1995.

Briscoe v. Reader's Digest, 483 P.2d 34, 36 (Calif. 1971).

Brown, Lisa. "Dead But Not Forgotten: Proposals for Imposing Liability for Defamation of
the Dead." *Texas Law Review* 67 (1989): 1525–1567.

Bruyneel, Kevin. "The King's Body: The Martin Luther King Jr. Memorial and the Politics
of Collective Memory." *History and Memory* 26 (2014): 75–108.

Buchanan, Elizabeth A. *Readings in Virtual Research Ethics: Issues and Controversies.*
Hershey, PA: Information Science Publications, 2004.

Bunker, Matthew D. "Takin' Care of Business: Confidentiality under the Business Exemption of the FOIA." *Public Relations Review* 21 (1995): 137–149.

Burke, Frank G. *Research and the Manuscript Tradition.* Lanham, MD: Scarecrow Press;
Society of American Archivists, 1997.

Burris, Scott, and Jen Welsh. "Regulatory Paradox: A Review of Enforcement Letters Issued by the Office for Human Research Protection." *Northwestern University Law Review* 101 (2007): 643–686.

Burton, Antoinette, ed. *Archive Stories: Facts, Fictions and the Writing of History.* Durham, SC: Duke University Press, 2005.

Butcher v. The Lincoln Journal, 2012 U.S. Dist. LEXIS 3883.

Cain, V.J.H. "The Ethics of Processing." *Provenance* 11 (1993): 39–55.

Cameron, Anderson, and Shirako Aiwa. "Are Individuals' Reputations Related to Their History of Behavior?" *Journal of Personality and Social Psychology* 94 (2008): 320–333.

Campbell, Richard T. "Risk and Harm Issues in Social Science Research." University of Urbana–Champaign, 2003. http://www.uiuc.edu\cas\cas_irb\.

Carp, E. Wayne. *Family Matters: Secrecy and Disclosure in the History of Adoption.* Cambridge, MA: Harvard University Press, 1998.

Cassidy v. Mark Nicolo, et al., 2005 U.S. Dist. LEXIS 34160.

Catsouras v. California Department of Highway Patrol, 181 Cal. App. 4th 856 (2010).

Chapman v. Journal Concepts, Inc., 528 F. Supp. 2d 1081 (D.Haw. 2007).

Charrow, Robert. "Censorship and Institutional Review Boards: Protection of Human Subjects: Is Expansive Regulation Counter-Productive?" *Northwestern University Law Review.* (2007): 707–718.

Cincinnati Enquirer v. Daniels, 844 N.E. 2d 1181 (Ohio 2006).

Cirasella, Jill Cariffe. "At Odds?: Archives and Privacy." *Current Studies in Librarianship* 24 (2000): 88–92.

Coe, Frederic L. "Censorship and Institutional Review Boards: The Costs and Benefits of a Well-Intended Parasite: A Witness and a Reporter on the IRB Phenomenon." *Northwestern University Law Review* 101 (2007): 723–733.

Cohen, Patricia. "As Ethics Panels Expand Grip, No Field Is Off Limits." *New York Times,* February 28, 2007, A15.

Columbia University. Augustus C. Long Health Sciences Library. "Access Policies: Access to Records Containing Protected Health Information." http://vesta.cumc.columbia.edu/library/archives/accesspatient.html.

Cook, Terry. "The Archive(s) Is A Foreign Country: Historians, Archivists, and the Changing Archival Landscape." *The American Archivist* 74 (2011): 600–632.

———. ed. *Controlling the Past.* Chicago, IL: Society of American Archivists, 2011.

Cook, Timothy E. *Freeing the Presses the First Amendment in Action.* Baton Rouge, LA: Louisiana State University Press, 2005.

Coté, William, and Roger Simpson, eds. *Covering Violence: A Guide to Ethical Reporting About Victims and Trauma.* New York: Columbia University Press, 2000.

Court of Justice of the European Union. Press Release no. 70/14, Luxembourg, May 13, 2014, at curia.europa.eu/jcms/jcms/P_127116.

Crenner, Christopher. "Race and Medical Practice in Kansas City's Free Dispensary." *Bulletin of the History of Medicine* 82 (2008): 820–846.

Danielson, Elena. *The Ethical Archivist.* Chicago, IL: Society of American Archivists, 2010.

d'Arouet, François-Marie. "Lettres à M. De Genonville. Lettre Premiere (1719)." *Œuvres Complete de Voltaire,* vol. 1. Paris: De L'Imprimerie de la Société Littéraire-Typographique, 1785.

Davis, Shelley L. *Unbridled Power: Inside the Secret Culture of the IRS.* New York: HarperBusiness, 1997.

Dayton, Cornelia Hughes. "Rethinking Agency, Recovering Voices." *American Historical Review* 109 (2004): 827–843.

Dean, Eric T., Jr. "Reflections on 'The Trauma of War' and *Shook Over Hell." Civil War History* 59 (2013): 414–418.

De Baets, Antoon. "A Declaration of the Responsibilities of Present Generations toward Past Generations." *History and Theory* 43 (2004): 130–164.

———. "Defamation Cases Against Historians," *History and Theory* 41 (2002): 346–366.

Department of Health and Human Services. "Advanced Notice of Proposed Rule Making on 45 CFR 46, 160, 164 Human Subjects Research Protections: Enhancing Protections for Research Subjects and Reducing Burden, Delay and Ambiguity for Investigators, Docket Number HSS-OPHS-2011–0005." July 26, 2011. http://www.regulations .gov/#!documentDetail;D=HHS-OPHS-2011–0005–0001.

Derrida, Jacque. *Archive Fever: A Freudian Interpretation.* Chicago, IL: University of Chicago Press, 1996.

Dunham, Chris. "Genealogy: Another Reason for Your Family to Hate You." *The Genealogue,* April 10, 2010. http://www.genealogue.com/2010_04_01_archive.html.

Duranti, Luciana. *Diplomatics: New Uses for an Old Science.* Lanham, MD: Society of American Archivists, Association of Canadian Archivists, The Scarecrow Press, 1998.

Emanuel, Ezekiel J., Lemmens Trudo, and Elliot Carl. "Should Society Allow Research Ethics to Be Run As For-Profit Enterprises?" *PLoS Medicine* 3 (2006). http://www.ncbi .nlm.nih.gov/pmc/articles/PMC1518668.

Fassin, Didier. "The Humanitarian Politics of Testimony: Subjectification through Trauma in the Israeli–Palestinian Conflict." *Cultural Anthropology* 23 (2008): 531–558.

"FAQ—IRB Website.doc." Institutional Review Board, Sponsored Programs and Research, Winthrop University. http://www2.winthrop.edu/spar/FAQ—IRB Website.doc.

Feldstein, Mark. "Kissing Cousins: Journalism and Oral History." *Oral History Review* 31 (2004): 1–22.

Fiske, Thomas. "Don't Dig Up the Past!" *GenealogyBlog,* April 13, 2010. Accessed July 13, 2014. http://www.genealogyblog.com/?p=8231.

Foreman, Gene. *The Ethical Journalist: Making Responsible Decisions in the Pursuit of News.* Chichester, UK; Malden, MA: Wiley-Blackwell, 2010.

Fox, Renee C., and Judith P. Swazey. "Medical Morality is not Bioethics—Medical Ethics in China and the United States." *Perspectives in Biology and Medicine* 27 (1984): 336–360.

Frankel, Mark S., and Sanyin Siang. "Ethical and Legal Aspects of Human Subjects Research on the Internet," a report to the American Association for the Advancement of Science (1999). http://www.aaas.org/page/shrl-ethics-law-activities.

Freud, Sigmund. *Reflections on War and Death.* Translated from the German by A. A. Brill and Alfred B. Kuttner. New York: Moffat, Yard and Co., 1918.

Gajda, Amy. "Judging Journalism: The Turn towards Privacy and Judicial Regulation of the Press." *California Law Review* 97, no. 2 (2009): 1039–1105.

Gates v. Discovery Communications, 101 P.3d 552 (Calif. 2004).

Gaudette, Marybeth. "Playing Fair with the Right to Privacy." *Archival Issues* 28 (2003–2004): 21–34.

George, Christine. "'Whatever You Say, You Say Nothing': Archives and the Belfast Project." Masters of Science in Information Studies, The University of Texas at Austin, 2012.

Geselbracht, Raymon H. "The Origins of Restrictions on Access to Personal Papers at the Library of Congress." *The American Archivist* 49 (1986): 142–162.

Gilbert v. Medical Economics, 665 F.2d 305, 306 (10th Cir. 1981).

Gold, Matea. "NBC Resolves Lawsuit over 'To Catch a Predator' Suicide." *LA Times*, June 24, 2008. http://latimesblogs.latimes.com/showtracker/2008/06/nbc-resolves-la.html.

Golden, Janet. *Message in a Bottle: The Making of Fetal Alcohol Syndrome.* Cambridge, MA: Harvard University Press, 2005.

Goldfarb, Ronald L. *In Confidence: When to Protect Secrecy and When to Require Disclosure.* New Haven, CT: Yale University Press, 2009.

Gonzaga University v. John Doe, 536 U.S. 273 (2002).

Goodwin, H. Eugene. *Groping for Ethics in Journalism.* Ames, IA: Iowa State University Press, 1983.

Greene, Mark A. "A Critique of Social Justice as an Archival Imperative: What *Is* It We're Doing That's All That Important?" *The American Archivist* 76 (2013): 302–334.

———. "Moderation in Everything, Access in Nothing? Opinions about Access Restrictions on Private Papers," *Archival Issues* 18 (1993): 31–41.

———. "MPLP: It's Not Just for Processing Anymore." *The American Archivist* 73 (2010): 175–203.

Greene, Mark A., and Dennis Meissner. "More Product, Less Process: Revamping Traditional Archival Processing." *The American Archivist* 68 (2005): 208–263.

Grossman, Eleanor L. "Privacy." *American Jurisprudence*, 2nd ed. Eagan, MN: Thomson Reuters, 2012.

"Guidance on IRB Review of Research Involving Existing Data Sets." The University of Chicago Social & Behavioral Sciences Institutional Review Board, 2007. http://humansubjects.uchicago.edu/sbrirb/publicpolicy.html.

Halstuk, Martin E. "Shielding Private Lives from Prying Eyes: The Escalating Conflict Between Constitutional Privacy and the Accountability Principle of Democracy." *CommLaw Conspectus* 11 (2003): 71–96.

———. "When Is Invasion of Privacy Unwarranted Under FOIA? An Analysis of the Supreme Court's 'Sufficient Reason' and 'Presumption of Legitimacy' Standards." *Journal of Law and Public Policy* 16 (2005): 361–399.

Hamburger, Philip. "Censorship and Institutional Review Boards: Getting Permission." *Northwestern University Law Review* 101 (2007): 466–473.

Hampsten, Elizabeth. *Read this Only to Yourself: The Private Writings of Midwestern Women, 1880–1910.* Bloomington, IN: Indiana University Press, 1982.

Harris, Verne. "The Archival Sliver: Power, Memory and Archives in South Africa." *Archival Science* 2 (2002): 63–86.

———. *Archives and Justice: A South African Perspective.* Chicago, IL: Society of American Archivists, 2007.

Haynes v. Alfred A. Knopf, Inc. and Nicolas Lemann, 8 F.3d 1222 (7th Cir. 1993).

Hemmer, Joseph J. *The Supreme Court and the First Amendment.* New York: Praeger, 1986.

Herrada, Julie. "Letters to the Unabomber: A Case Study and Some Reflections." *Archival Issues* 28 (2003–2004): 35–46.

Hoffman, Sharona, and Jessica Wilen Berg. "The Suitability of IRB Liability." *University of Pittsburg Law Review* 67 (2005): 365–427.

Hoffmaster, Barry, ed. *Bioethics in Social Context.* Philadelphia: Temple University

Press, 2001.

———. "Can Ethnography Save the Life of Medical Ethics?" *Social Science & Medicine* 35 (1992): 1421–1431.

Hoff-Wilson, Joan. "Access to Restricted Collections: The Responsibility of Professional Historians." *The American Archivist* 46 (1983): 441–448.

Holm, Søren. "The Privacy of Tutankhamen—Utilizing the Genetic Information in Stored Tissue Samples." *Theoretical Medicine* 22 (2001): 437–449.

hooks, bell. "Marginality as a Site of Resistance." In *Out There: Marginalization and Contemporary Cultures,* edited by Russell Ferguson, et al., 241-243. Cambridge, MA: MIT Press, 1990.

Howden, Lindsay M., and Julie A. Meyer. "Age and Sex Composition: 2010," *2010 Census Briefs.* Washington, DC: United States Census Bureau, May 2011.

Hulteng, John L. *The Messenger's Motives: Ethical Problems of the News Media.* Englewood Cliffs, NJ: Prentice-Hall, 1976.

"Human Subjects Research Protections: Enhancing Protections for Research Subjects and Reducing Burden, Delay and Ambiguity for Investigators." http://www.regulations .gov/#!documentDetail;D=HHS-OPHS-2011-0005-0001.

Iacovina, Livia, and Todd Malcolm. "The Long-Term Preservation of Identifiable Personal Data: A Comparative Archival Perspective on Privacy Regulatory Models in the European Union, Australia, Canada and the United States." *Archival Science* 7 (2007): 107–127.

Imber-Black, Evan, ed. *Secrets in Families and Family Therapy.* 1st ed. New York: Norton, 1993.

International Council on Archives. *Memory of the World at Risk: Archives Destroyed, Archives Reconstituted. Archivum,* vol. XLII. Munich: K. G. Saur, 1996.

Iraymi, Raymond. "Give the Dead Their Day in Court: Implying a Private Cause of Action for Defamation of the Dead from Criminal Libel Statutes." *Fordham Intellectual Property, Media & Entertainment Law Journal* 9 (1999): 1083–1124.

Irons, et al. v. Federal Bureau of Investigation, 880 F.2d 1446 (1989).

Jacquette, Dale. *Journalistic Ethics: Moral Responsibility in the Media.* Upper Saddle River, NJ: Pearson Prentice Hall, 2007.

Janesick, Valerie J. *Oral History for the Qualitative Researcher: Choreographing the Story.* New York: Guilford Press, 2010.

Jerome, Fred. *The Einstein File: J. Edgar Hoover's Secret War Against the World's Most Famous Scientist.* New York: St. Martin's Griffin, 2002.

Jimerson, Randall C., ed. *American Archival Studies: Readings in Theory and Practice.* Chicago, IL: Society of American Archivists, 2000.

———. "Archivists and Social Responsibility: A Response to Mark Greene." *The American Archivist* 76 (2013): 335–345.

Justice v. Belo Broadcasting Corporation, 472 F. Supp. 145 (N.D. Tex. 1979).

Kaplan, Diane E. "The Stanley Milgram Papers: A Case Study on Appraisal of and Access to Confidential Data Files." *The American Archivist* 59 (1996): 288–297.

Keeton, Page, and William L. Prosser, eds. *Prosser and Keeton on the Law of Torts.* 5th ed., student ed., Hornbook Series. St. Paul, MN: West, 1984.

Kels, Maj. Charles G. "Privacy after Death?" *Reporter* 38 (2011): 36–40.

Kern, Steven. *The Culture of Love: Victorians to Moderns.* Cambridge, MA: Harvard University Press, 1992.

Kleinman, Arthur, ed. *Writing at the Margin: Discourse Between Anthropology and Medi-*

cine. Berkeley: University of California Press, 1995.

Kleinman, Arthur, Renée C. Fox, and Allan M. Brandt, eds. *Bioethics and Beyond, Daedalus* 128 (1999), 1–326.

Laurie, Graeme T. *Genetic Privacy: A Challenge to Medico-Legal Norms.* Cambridge; New York: Cambridge University Press, 2002.

Lawrence, Susan C. "Access Anxiety: HIPAA and Historical Research." *Journal of the History of Medicine and Allied Sciences* 62, no. 4 (2007): 422–460.

Lee v. Weston, 402 N.E.2d 23 (Ind. App. 1980).

Letocha, Phoebe Evans, and Emily R. Novak Gustainis. "Recommended Practices for Enabling Access to Manuscript and Archival Collections Containing Health Information about Individuals." http://www.medicalheritage.org/2015/02/now-available-recommended-practices-for-enabling-access-to-manuscript-and-archival-collections-containing-health-information-about-individuals/.

Lilly, J. Robert. *Taken By Force: Rape and American GIs in Europe During World War II.* Houndsmills, Hampshire, UK: Palgrave Macmillan, 2007.

Linenthal, Edward T., and Tom Engelhardt. *History Wars: The Enola Gay and Other Battles for the American Past.* New York: Henry Holt, 1996.

Lingren, James, Dennis Murashko, and Matthew R. Ford, eds. "Special Issue: Symposium on Censorship and Institutional Review Boards." *Northwestern University Law Review* 101, no. 2 (2007).

Livelton, Trevor. *Archival Theory, Records and the Public.* Lanham, MD: The Society of American Archivists and the Scarecrow Press, 1996.

Loftus, Elizabeth F., and Melvin J. Guyer. "Who Abused Jane Doe? The Hazards of the Single Case History, Part I and Part II." *The Skeptical Inquirer* 26 (2002): 24–32, 37–42.

Logan v. Dept. of Veterans Affairs, 357 F. Supp. 2d 149, 155 (D.D.C. 2004).

Lopez, Jose. "How Sociology Can Save Bioethics . . . Maybe." *Sociology of Health & Illness* 26 (2004): 875–896.

Mackowski, Philip A. *Diagnosing Giants: Solving the Medical Mysteries of Thirteen Patients Who Changed the World.* New York: Oxford University Press, 2013.

MacNeil, Heather. *Without Consent: The Ethics of Disclosing Personal Information in Public Archives.* Metuchen, NJ: The Scarecrow Press and the Society of American Archivists, 1992.

Macri, Martha, and James Sarmento. "Respecting Privacy: Ethical and Pragmatic Considerations." *Language & Communication* 30, no. 3 (2010): 192–197.

Mallon, Thomas. *Stolen Words: The Classic Book on Plagiarism.* 2nd ed. San Diego, CA: Harcourt, Inc., 2001.

Margalit, Avishai. *The Ethics of Memory.* Cambridge, MA: Harvard University Press, 2002.

Markowitz, Gerald E., and David Rosner. *Deceit and Denial: The Deadly Politics of Industrial Pollution.* Berkeley: University of California Press, 2002.

Masterton, Malin, Gert Helgesson, Anna T. Höglund, and Mats G. Hansson. "Queen Christiana's Moral Claim on the Living: Justification of a Tenacious Moral Intuition." *Medicine, Health Care and Philosophy* 10 (2007): 321–327.

McKemmish, Sue, Michael Piggott, Barbara Reed, and Frank Upward, eds. *Archives: Recordkeeping in Society.* Wagga Wagga, New South Wales: Center for Information Studies, 2005.

McMurtrie, Beth. "Secrets from Belfast. How Boston College's Oral History of the Troubles Fell Victim to an International Murder Investigation." *The Chronicle of Higher*

Education 60 (January 31, 2014).

Meeropol v. Nizer, 417 F. Supp. 1201 (S.D. New York 1976).

Megill, Allan. *Historical Knowledge, Historical Error: A Contemporary Guide to Practice.* Chicago, IL: University of Chicago Press, 2007.

Melanson, Philip H. *Secrecy Wars: National Security, Privacy and the Public's Right to Know.* Washington, DC: Brassey's Inc., 2001.

Mencken, Jennifer A. "Supervising Secrecy: Preventing Abuses Within Bank Secrecy and Financial Privacy Systems." *Boston College International and Comparative Law Review* 21 (1998): art. 5.

Metcalfe, Wayne J., and Melvin P. Thatcher. "Serving the Genealogical and Historical Research Communities: An Overview of Records Access and Data Privacy Issues." *World Library and Information Congress: 74th IFLA General Conference and Council, August 10–14, 2008.* http://archive.ifla.org/IV/ifla74/papers/117-Metcalfe_Thatcher -en.pdf.

Meyer, Philip. *Ethical Journalism: A Guide for Students, Practitioners, and Consumers.* Communications. New York: Longman, 1987.

Meyers, Christopher, ed. *Journalism Ethics: A Philosophical Approach.* New York: Oxford University Press, 2010.

Middlebrook, Diane Wood. *Anne Sexton: A Biography.* Boston: Houghton Mifflin, 1991.

———. "Telling Secrets." In *The Seductions of Biography*, edited by Mary Rhiel and David Suchoff, 124–129. New York: Routledge, 1996.

Miller, Harold L. "Will Access Restrictions Hold Up in Court? The FBI's Attempt to Use the Braden Papers at the State Historical Society of Wisconsin." *The American Archivist* 52 (1983): 180–190.

Minnesota History Center. "Application to Restricted Records in State Archives." http:// sites.mnhs.org/library/content/faq-restricted-collections.

National Archives and Records Administration v. Favish, 541 U.S. 157, 163 (2004)."National Center for Education Statistics." U.S. Department of Education. Accessed August 31, 2011. http://nces.ed.gov.

National Commission for the Protection of Human Subjects of Biomedical and Behavioral Research. *Belmont Report: Ethical Principles and Guidelines for the Protection of Human Subjects.* 1978. http://ohsr.od.nih.gov/guidelines/belmont.html.

National Research Council. *Expanding Access to Research Data: Reconciling Risks and Opportunities.* Washington, DC: National Academies Press, 2005.

Neuenschwander, John A. *A Guide To Oral History and The Law.* New York: Oxford University Press, 2009.

New York Times Company. *"New York Times* Company Policy on Ethics in Journalism." http://www.nytco.com/press/ethics.html.

Novick, Peter. *That Noble Dream: The "Objectivity Question" and the American Historical Profession.* New York: Cambridge University Press, 1988.

Office of Human Research Protections. Secretary's Advisory Committee on Human Research Protections. "Considerations and Recommendations Concerning Internet Research and Human Subjects Research." http://www.hss-gov/ohrp/sachrp/commsec/.

Ohm, Paul. "Broken Promises of Privacy: Responding to the Surprising Failure of Anonymization." *UCLA Law Review* 57 (2009–2010): 1701–1777.

Olson, Marilyn. "'Halt, Blind, Lame, Sick and Lazy': Care of the Poor in Cedar County, Iowa." *The Annals of Iowa* 69 (2010): 131–172.

Partridge, E. "Posthumous Ethics and Posthumous Respect." *Ethics* 91 (1981): 243–264.

Phelan, James. *Living to Tell About It: A Rhetoric and Ethics of Character Narration.* Ithaca, NY: Cornell University Press, 2005.

———. *Narrative as Rhetoric: Technique, Audiences, Ethics, Ideology.* Columbus, OH: Ohio State University Press, 1996.

———. ed. *Reading Narrative: Form, Ethics, Ideology.* Columbus, OH: Ohio State University Press, 1989.

Potter, Clare Bond, and Renee C. Romano, eds. *Doing Recent History.* Athens, GA: The University of Georgia Press, 2012.

Pressman, Jack. *Last Resort: Psychosurgery and the Limits of Medicine.* San Francisco, CA: University of California Press, 1998.

"Privacy of School Records Laws." *Law Library: American Law and Legal Information— State Laws and Statutes,* July 10, 2011. http://law.jrank.org/pages/11819/ Privacy-School-Records.

Privacy Rights Clearing House: Empowering Consumers, Protecting Privacy. "Public Records on the Internet: The Privacy Dilemma," (2002). https://www.privacyrights.org/ ar/onlinepubrecs.htm.

Prosser, William, and Page Keeton, *Handbook of the Law of Torts,* 5th ed. St. Paul, MN: Thomson/West, 1984.

"Publication of Private Facts." *Digital Media Law Project,* June 28, 2012. http://www .citmedialaw.org/subject-area/publication-private-facts.

Reagan, Leslie. *When Abortion was a Crime: Women, Medicine and Law in the United States, 1867–1973.* Berkeley: University of California Press, 1997.

Resnick, David B., and Richard R. Sharp. "Protecting Third Parties in Human Subjects Research." *IRB: Ethics and Human Research* 28 (2006): 1–7.

Reverby, Susan. *Examining Tuskegee: The Infamous Syphilis Study and its Legacy.* Chapel Hill, NC: University of North Carolina Press, 2009.

Richards, Robert D., and Clay Calvert. "Suing the Media, Supporting the First Amendment: The Paradox of Neville Johnson and the Battle for Privacy." *Albany Law Review* 67 (2004): 1097–1135.

Ritchie, Donald A. *Doing Oral History.* New York: Twayne, 1995.

Roberts, Alasdair. *Blacked Out: Government Secrecy in the Information Age.* Cambridge; New York: Cambridge University Press, 2006.

Roberts, Alexa. "Trust Me, I Work for the Government: Confidentiality and Public Access to Sensitive Information." *American Indian Quarterly* 25 (2001): 13–17.

Robin, Ron Theodore. *Scandals and Scoundrels: Seven Cases That Shook the Academy.* Berkeley: University of California Press, 2004.

Robine, J. M., and M. Allard. "Jeanne Calement: Validation of the Duration of Her Life." In *Validation of Exceptional Longevity.* Odense Monographs on Population Aging 8, edited by B. Jeune and J. W. Vaupel, 145–172. Odense, Denmark: Odense University Press, 1999.

Rosen, Jeffrey. "The Right to Be Forgotten." *Stanford Law Review Online* 16 (2012): 88–92. http://www.stanfordlawreview.org/online/privacy-paradox/right-to-be-forgotten.

Roshto v. Hebert, et al., 439 SO.2d 428, 431 (Louis. 1983).

Rosler, Hannes. "Dignitarian Posthumous Personality Rights—An Analysis of U.S. and German Constitutional and Tort Law." *Berkeley Journal of International Law* 26, no. 1 (2008): 153–205.

Roth, Randolph. *American Homicide.* Cambridge, MA: Belknap Press of Harvard University Press, 2009.

Rubel, Dejah T. "Accessing their Voices from Anywhere: Analysis of the Legal Issues Surrounding the Online Use of Oral Histories." *Archival Issues* 31 (2007): 171–187.

Rule, John C., and Ben S. Trotter. *A World of Paper: Louis XIV, Colbert de Tracy and the Rise of the Information State.* Montreal: McGill-Queen's University Press, 2014.

Sachdev, Paul. *Unlocking the Adoption Files.* Lexington, MA: Lexington Books, 1989.

Savala v. Freedom Communications, 2006 Cal. App. Unpub. LEXIS 5609.

Schaffer, Kay, and Sidonie Smith, eds. *Human Rights and Narrated Lives: The Ethics of Recognition.* New York: Palgrave Macmillan, 2004.

Schrag, Zachary M. *Ethical Imperialism: Institutional Review Boards and the Social Sciences, 1965–2007.* Baltimore, MD: Johns Hopkins University Press, 2010.

———. "Institutional Review Blog: News and Commentary about Institutional Review Board Oversight of the Humanities and Social Sciences." http://www.institutionalreviewblog.com/.

Schrecker v. Department of Justice, 349 F.3d 657 (D.C. Cir. 2003).

Schwarz, Judith. "The Archivist's Balancing Act: Helping Researchers While Protecting Individual Privacy." *The Journal of American History* 79 (1992): 183–184.

Scott, James C. *Domination and the Arts of Resistance: Hidden Transcripts.* New Haven, CT: Yale University Press, 1990.

Seattle Times Co. v. Rhinehart, et al., 467 U.S. 20 (1984).

Seeman, Dean. "Naming Names: The Ethics of Identification in Digital Library Metadata." *Knowledge Organization* 39 (2012): 325–331.

Seidler, Victor Jeleniewski. "Obligations to the Dead: Historical Justice and Cultural Memory." *European Judaism* 46 (2013): 12–31.

Shapira, Anita. "The Holocaust: Private Memories, Public Memory." *Jewish Social Studies* n.s. 4 (1998): 40–58.

Shaw, David Gary. "Happy in Our Chains? Agency and Language in the Postmodern Age." *History and Theory* 40 (2001): 1–9.

Sheftel, Anna, and Stacey Zembrzycki. "Only Human: A Reflection on the Ethical and Methodological Challenges of Working with 'Difficult' Stories." *The Oral History Review* 37 (2010): 191–214.

Shopes, Linda. "Human Subjects and IRB Review." *Oral History Association*, n.d. http://www.oralhistory.org/do-oral-history/oral-history-and-irb-review.

Sieber, Joan E. "Issues Presented by Mandatory Reporting Requirements to Researchers of Child Abuse and Neglect." *Ethics Behavior* 4 (1994): 1–22.

———. ed. *NIH Readings on the Protections of Human Subjects in Behavioral and Social Science Research.* Frederick, MD: University Publications of America, 1984.

Silverman, Justin. "The Catsouras Photos: Will a Family's Privacy Interest Impede Press Access?" *Citizen Media Law Project*, February 11, 2010. http://www.citmedialaw.org/blog/2010/catsouras-photos-will-familys-privacy-interest-impede-press-access.

Simms, Karl, ed. *Ethics and the Subject. Critical Studies,* vol. 8. Amsterdam: Rodopi, 1997.

Simon, Roger I. *The Touch of the Past: Remembrance, Learning, Ethics.* New York: Palgrave Macmillan, 2005.

Simpson, Roger. *Covering Violence: A Guide to Ethical Reporting about Victims and Trauma.* 2nd ed. New York: Columbia University Press, 2006.

Smart, Carol. "Family Secrets: Law and Understandings of Openness in Everyday Relationships." *Journal of Social Policy* 38 (2009): 551–567.

Smith, Janna Malamud. *Private Matters: In Defense of the Personal Life.* Reading, MA:

Addison-Wesley, 1997.

Smolensky, Kirsten Rabe. "Rights of the Dead." *Hofstra Law Review* 37 (2009): 763–803.

Society of American Archivists. "SAA Core Values Statement and Code of Ethics." 2011–2012. http://www2.archivists.org/statements/saa-core-values-statement-and-code-of-ethics.

Solove, Daniel J. *The Digital Person: Technology and Privacy in the Information Age.* New York: New York University Press, 2004.

———. *The Future of Reputation: Gossip, Rumor, and Privacy on the Internet.* New Haven, CT: Yale University Press, 2007.

Sommer, Barbara W., and Mary Key Quinlin. *The Oral History Manual.* New York: AltaMira Press, 2002.

Sperling, Daniel. *Posthumous Interests: Legal and Ethical Perspectives.* New York: Cambridge University Press, 2008.

Stark, Laura. *Behind Closed Doors: IRBs and the Making of Ethical Research.* Chicago, IL: University of Chicago Press, 2012.

State of Nebraska, ex rel. Adams County Historical Society vs. Nancy Kinyoun, 277 Neb. 749 (2009).

Steele, Meili. *Theorizing Textual Subjects: Agency and Oppression. Literature, Culture, Theory* vol. 21. New York: Cambridge University Press, 1997.

Stephens, Martha. *The Treatment: The Story of Those Who Died in the Cincinnati Radiation Tests.* Durham, NC: Duke University Press, 2002.

Stoddard, Daniel G. "Falling Short of Fundamental Fairness: Why Institutional Review Board Regulations Fail to Provide Procedural Due Process." *Creighton University Law Review* 43 (2010): 1275–1327.

Stoler, Ann Laura. *Along the Archival Grain: Epistemic Anxieties and Colonial Common Sense.* Princeton, NJ: Princeton University Press, 2009.

Strous, R. D. "To Protect or To Publish: Confidentiality and the Fate of the Mentally Ill Victims of Nazi Euthanasia." *Journal of Medical Ethics* 35 (2009): 361–364.

Tauris, Carol. "The High Cost of Skepticism." *Skeptical Inquirer*, no. 26 (2002). http://www.csicop.org/si/show/high_cost_of_skepticism.

The Johns Hopkins Medical Institutions. Privacy Board. "Application for a Waiver of Authorization for Research Use or Disclosure of Protected Health Information (PHI) and Other Personal Information that is Protected by Law." http://www.medicalarchives.jhmi.edu/hipaaform.html.

"The Laws in Your State." *RAINN: Rape, Abuse & Incest National Network Website*, n.d. http://www.rainn.org/public-policy/laws-in-your-state.

Todd, Malcolm. "Power, Identity, Integrity, Authenticity, and the Archives: A Comparative Study of the Application of Archival Methodology to Contemporary Privacy." *Archivaria* 61 (2006): 181–215.

Townsend, Robert. "Oral History and Review Boards: Little Gain and More Pain." *AHA Perspectives*, 2006. http://www.historians.org/perspectives/issues/2006/0602/0602new1.cfm.

Tuniok, Mark. "Privacy and Punishment." *Social Theory and Practice* 39 (2013): 643–668.

United States of America v. David S. Chase, M.D., 2005 U.S. Dist. LEXIS 38676.

United States Census Bureau. *Age and Sex Composition: 2010* (May 2011).

Vangelisti, Anita, and P. Caughlin John. "Revealing Family Secrets: The Influence of Topic, Function, and Relationships." *Journal of Social and Personal Relationships* 14, no.

5 (1997): 679–705.

Volokh, Eugene. "Freedom of Speech and Information Privacy: The Troubling Implications of a Right to Stop People from Speaking About You." *Stanford Law Review* 52 (2000): 1049–1124.

Wallenstein, Peter. *Tell the Court I Love My Wife: Race, Marriage, and Law—An American History.* New York: Palgrave Macmillan, 2002.

Walther, Joseph B. "Research Ethics in Internet-Enabled Research: Human Subjects Issues and Methodological Myopia." *Ethics and Information Technology* 4 (2002): 205–212.

Ward, Stephen J. A. *The Invention of Journalism Ethics the Path to Objectivity and Beyond.* Montreal: McGill-Queen's University Press, 2006.

Warren, Samuel D., and Louis D. Brandeis. "The Right to Privacy." *Harvard Law Review* 4 (1890): 193–200.

Warshauer, Matthew, and Michael Sturges. "Difficult Hunting: Accessing Connecticut Patient Records of Learn About Post-Traumatic Stress Disorder during the Civil War." *Civil War History* 59 (2013): 419–452.

Watson, Thomas E. *Life and Times of Thomas Jefferson.* New York: D. Appleton and Co., 1903.

Way, L. Randall. "ORS and IRB Investigation and Punishment of Alleged Faculty Misconduct." *The Faculty Advocate* 8 (2008). http://cas.umkc.edu/aaup/facadv23.html#Wray.

Weaver, Gina Marie. *Ideologies of Forgetting: Rape in the Vietnam War.* Albany, NY: State University of New York Press, 2010.

Westerman, Casey S. "Last Words: Suicide Notes, Ownership, Access and Privacy." Paper presented at the annual meeting of the Society of American Archivists, Washington, DC, August 15, 2014.

Wexler, Alice. *The Woman Who Walked into the Sea: Huntington's and the Making of a Genetic Disease.* New Haven, CT: Yale University Press, 2008.

White, Ronald F. "Institutional Review Board Mission Creep: The Common Rule, Social Science, and the Nanny State." *The Independent Review* 11 (2007): 547–564.

Whiteman, Natasha. *Undoing Ethics: Rethinking Practice in Online Research.* New York: Springer, 2012.

Wiener, Jon. *Historians in Trouble: Plagiarism, Fraud, and Politics in the Ivory Tower.* New York: New Press; Distributed by W. W. Norton, 2005.

Williams, Peter. "Is Financial Privacy Preventing Legitimate Research?" *I/S: A Journal of Law & Policy for the Information Society* 5 (2010): 555–567.

Woodage, Trevor. "Relative Futility: Limits to Genetic Privacy Protection Because of the Inability to Prevent Disclosure of Genetic Information by Relatives." *Minnesota Law Review* 95 (2010): 682–713.

Wrathall, John D. "Provenance as Text: Reading the Silences around Sexuality in Manuscript Collections." *The Journal of American History* 79 (1992): 156–178.

Yaco, Sonia. "Balancing Privacy and Access in School Desegregation Collections: A Case Study." *The American Archivist* 73 (2010): 637–668.

Young v. That Was The Week That Was, 312 F. Supp. 1337 (N.D. Ohio 1969).

Index

About the Author

Susan C. Lawrence is an associate professor of history at the Ohio State University. She received her BA in mathematics at Pomona College and her PhD in the history of medicine at the Institute for the History and Philosophy of Science and Technology at the University of Toronto.

Available titles in the Critical Issues in Health and Medicine series:

the Feingold Diet

Rosemary A. Stevens, Charles E. Rosenberg, and Lawton R. Burns, eds., *History and Health Policy in the United States: Putting the Past Back In*

Barbra Mann Wall, *American Catholic Hospitals: A Century of Changing Markets and Missions*

Frances Ward, *The Door of Last Resort: Memoirs of a Nurse Practitioner*

Printed and bound by CPI Group (UK) Ltd, Croydon, CR0 4YY

30/09/2024

14566217-0001